Digital Electronics: A Practical Approach

Digital Electronics: A Practical Approach

Kate Timberlake

STATES
ACADEMIC PRESS
www.statesacademicpress.com

Digital Electronics: A Practical Approach
Kate Timberlake
ISBN: 978-1-63989-154-2 (Hardback)

STATES
ACADEMIC PRESS

Published by States Academic Press,
109 South 5th Street,
Brooklyn, NY 11249, USA

Cataloging-in-Publication Data

Digital electronics : a practical approach / Kate Timberlake.
p. cm.
Includes bibliographical references and index.
ISBN 978-1-63989-154-2
1. Digital electronics. 2. Electronic systems. 3. Electronics.
I. Timberlake, Kate.
TK7868.D5 D54 2022
621.381--dc23

For more information regarding States Academic Press and its products, please visit the publisher's website www.statesacademicpress.com

Table of Contents

Table of Contents

Preface

The field of electronics that is concerned with the engineering devices which make use of or produce digital signals is known as digital electronics. The electrical circuits used in such devices are large assemblies of logic gates which are packed into integrated circuits. The logic families can be broadly categorized into resistor-transistor logic, diode-transistor logic, transistor-transistor logic, emitter coupled logic and CMOS logic. Signals produced using digital systems can be transmitted without any significant degradation. They also provide easier information storage and access methods. Some of the important aspects of designing a digital circuit are representations, synchronous systems, asynchronous systems, computer architecture, design tools and register transfer systems. This book presents the complex subject of digital electronics in the most comprehensible and easy to understand language. Different approaches, evaluations and methodologies and advanced studies on digital electronics have been included herein. Those in search of information to further their knowledge will be greatly assisted by this book.

A detailed account of the significant topics covered in this book is provided below:

Chapter 1- A signal that represents data as a sequence of discrete values is termed as a digital signal. There are different types of circuits such as digital, arithmetic, digital integrated, combinational and sequential circuits. This chapter has been carefully written to provide an easy understanding of various concepts related to digital signals such as comparator, finite state machine as well as these types of circuits.

Chapter 2- The collection of different integrated circuit chips which perform different logic gate functions like AND, OR, NOT but have similar input, output and integrated circuit characteristics are termed as logic families. A few of the major types of logic families are resistor-transistor logic, diode-transistor logic, emitter-coupled logic, transistor-transistor logic and MOS logic family. This chapter will explain briefly all these divisions of logic families.

Chapter 3- Some of the different categories of logic gates are NOT, AND, OR, NAND, NOR, XOR and XNOR gates. The aim of this chapter is to explore different types of logic gates which are closely related to its progress. These topics are crucial for a complete understanding of logic gates.

Chapter 4- The system of naming numbers is known as number system. The different bases of number system in digital electronics are the decimal, binary, octal, and hexadecimal number system. The chapter discusses the key types of number systems such as binary number system, binary code, binary-coded decimal, bitwise operation and Boolean algebra to provide an extensive understanding of the subject.

Chapter 5- Some of the concepts which are studied within digital electronics are: delays and hazards, Boolean algebra, Karnaugh maps, encoder, decoder, multiplexers and demultiplexers, Arithmetic Logic Unit (ALU) and digital system design. This chapter discusses in detail all these concepts.

I would like to make a special mention of my publisher who considered me worthy of this opportunity and also supported me throughout the process. I would also like to thank the editing team at the back-end who extended their help whenever required.

Kate Timberlake

1

Introduction to Digital Electronics

A signal that represents data as a sequence of discrete values is termed as a digital signal. There are different types of circuits such as digital, arithmetic, digital integrated, combinational and sequential circuits. This chapter has been carefully written to provide an easy understanding of various concepts related to digital signals such as comparator, finite state machine as well as these types of circuits.

Digital electronics are those electronics systems that use a digital signal instead of an analog signal. Digital electronics are the most common representation of Boolean algebra and are the basis of all digital circuits for computers, mobile phones, and numerous other consumer products. The most common fundamental unit of digital electronics is the logic gate. By combining numerous logic gates (from tens to hundreds of thousands) more complex systems can be created. The complex system of digital electronics is collectively referred to as a digital circuit.

A digital circuit that acts as a binary clock, hand-wired on a series of breadboards.

To most electronic engineers, the terms "digital circuit", "digital system" and "logic" are interchangeable in the context of digital circuits. The usual advantages of digital circuits when compared to analog circuits are:

- Digital systems interface well with computers and are easy to control with software. It is often possible to add new features to a digital system without changing hardware, and to do this remotely, just by uploading new software. Design errors or bugs can be worked-around with a software upgrade, after the product is in customer hands.
- Information storage can be much easier in digital systems than in analog ones. In particular, the great noise-immunity of digital systems makes it possible to store data and retrieve it later without degradation. In an analog system, aging and wear and tear will degrade the information in storage, but in a digital system, as long as the wear and tear is below a certain level, the information can be recovered perfectly.

Robustness

One of the primary advantages of digital electronic stems from it simply being digital: robustness. The robustness stems from the fact that if the noise is less than the noise margin then the system performs as if there were no noise at all. This is a necessary and sufficient property for a circuit to be considered digital. However, if the noise exceeds this level, the circuit can give unexpected and undesired results — perhaps catastrophic incorrect results.

Digital signals can thus be regenerated to achieve lossless data transmission, within certain limits. Analog signal transmission and processing, by contrast, always introduces noise. Theoretically, there is no data-loss when copying digital data. This is a great advantage over analog systems, which faithfully reproduce every bit of noise that makes its way into the signal.

Disadvantages

Digital circuits use more energy than analog circuits to accomplish the same calculations and signal processing tasks, thus producing more heat as well. In portable or battery-powered systems this can be a major limiting factor, but in a situation where power is plentiful, a digital system is often preferred because of all the advantages listed above, especially that of (re-)programmability and ease of upgrading without requiring hardware changes.

A particular example is the cellular telephone, which being a battery-powered portable device, uses a low-power analog front-end to acquire and tune in the radio signal from the base station. The base station, being in a fixed location with access to the power grid, can afford to use power-hungry software-defined (digital) radio techniques that digitize the signal essentially at the antenna (after wideband filtering and down conversion to intermediate frequency) and performs all channelization and demodulation via software-driven calculations. Such base stations can be reprogrammed, potentially via remote control, to process the signals used in future cellular standards as those standards become available.

Digital circuits are sometimes more expensive, especially in small quantities. The world in which we live is analog, and signals from this world such as light, temperature, sound, electrical conductivity, electric and magnetic fields, and phenomena such as the flow of time, are for most practical purposes continuous and thus analog quantities rather than discrete digital ones. For a digital system to do useful things in the real world, translation from the continuous realm to the discrete digital realm must occur, resulting in quantization errors. This problem can usually be mitigated by designing the system to store enough digital data to represent the signal to the desired degree of fidelity. The Nyquist-Shannon sampling theorem provides an important guideline as to how much digital data is needed to accurately portray a given analog signal.

Fragility

Digital systems can be fragile, in that if a single piece of digital data is lost or misinterpreted, the meaning of large blocks of related data can completely change. This problem can be mitigated by designing the digital system for robustness. For example, a parity bit or other error-detecting or error-correcting code can be inserted into the signal path so that if less than a certain fraction of the data is corrupted, the system can determine that this has occurred and possibly uncorrupt the

1

Introduction to Digital Electronics

A signal that represents data as a sequence of discrete values is termed as a digital signal. There are different types of circuits such as digital, arithmetic, digital integrated, combinational and sequential circuits. This chapter has been carefully written to provide an easy understanding of various concepts related to digital signals such as comparator, finite state machine as well as these types of circuits.

Digital electronics are those electronics systems that use a digital signal instead of an analog signal. Digital electronics are the most common representation of Boolean algebra and are the basis of all digital circuits for computers, mobile phones, and numerous other consumer products. The most common fundamental unit of digital electronics is the logic gate. By combining numerous logic gates (from tens to hundreds of thousands) more complex systems can be created. The complex system of digital electronics is collectively referred to as a digital circuit.

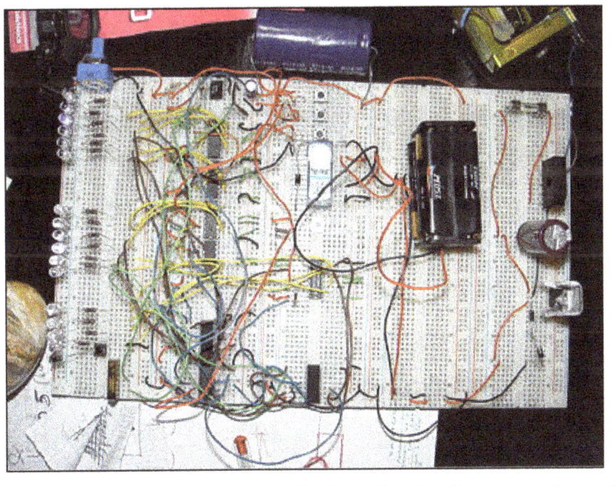

A digital circuit that acts as a binary clock, hand-wired on a series of breadboards.

To most electronic engineers, the terms "digital circuit", "digital system" and "logic" are interchangeable in the context of digital circuits. The usual advantages of digital circuits when compared to analog circuits are:

- Digital systems interface well with computers and are easy to control with software. It is often possible to add new features to a digital system without changing hardware, and to do this remotely, just by uploading new software. Design errors or bugs can be worked-around with a software upgrade, after the product is in customer hands.

- Information storage can be much easier in digital systems than in analog ones. In particular, the great noise-immunity of digital systems makes it possible to store data and retrieve it later without degradation. In an analog system, aging and wear and tear will degrade the information in storage, but in a digital system, as long as the wear and tear is below a certain level, the information can be recovered perfectly.

Robustness

One of the primary advantages of digital electronic stems from it simply being digital: robustness. The robustness stems from the fact that if the noise is less than the noise margin then the system performs as if there were no noise at all. This is a necessary and sufficient property for a circuit to be considered digital. However, if the noise exceeds this level, the circuit can give unexpected and undesired results — perhaps catastrophic incorrect results.

Digital signals can thus be regenerated to achieve lossless data transmission, within certain limits. Analog signal transmission and processing, by contrast, always introduces noise. Theoretically, there is no data-loss when copying digital data. This is a great advantage over analog systems, which faithfully reproduce every bit of noise that makes its way into the signal.

Disadvantages

Digital circuits use more energy than analog circuits to accomplish the same calculations and signal processing tasks, thus producing more heat as well. In portable or battery-powered systems this can be a major limiting factor, but in a situation where power is plentiful, a digital system is often preferred because of all the advantages listed above, especially that of (re-)programmability and ease of upgrading without requiring hardware changes.

A particular example is the cellular telephone, which being a battery-powered portable device, uses a low-power analog front-end to acquire and tune in the radio signal from the base station. The base station, being in a fixed location with access to the power grid, can afford to use power-hungry software-defined (digital) radio techniques that digitize the signal essentially at the antenna (after wideband filtering and down conversion to intermediate frequency) and performs all channelization and demodulation via software-driven calculations. Such base stations can be reprogrammed, potentially via remote control, to process the signals used in future cellular standards as those standards become available.

Digital circuits are sometimes more expensive, especially in small quantities. The world in which we live is analog, and signals from this world such as light, temperature, sound, electrical conductivity, electric and magnetic fields, and phenomena such as the flow of time, are for most practical purposes continuous and thus analog quantities rather than discrete digital ones. For a digital system to do useful things in the real world, translation from the continuous realm to the discrete digital realm must occur, resulting in quantization errors. This problem can usually be mitigated by designing the system to store enough digital data to represent the signal to the desired degree of fidelity. The Nyquist-Shannon sampling theorem provides an important guideline as to how much digital data is needed to accurately portray a given analog signal.

Fragility

Digital systems can be fragile, in that if a single piece of digital data is lost or misinterpreted, the meaning of large blocks of related data can completely change. This problem can be mitigated by designing the digital system for robustness. For example, a parity bit or other error-detecting or error-correcting code can be inserted into the signal path so that if less than a certain fraction of the data is corrupted, the system can determine that this has occurred and possibly uncorrupt the data, or ask for the corrupted data to be resent. In a state-machine, the state transition logic can be designed to catch all unused states and trigger a reset sequence or other error recovery routine. For example, it is standard practice in embedded software design to fill unused program memory with interrupt instructions that point to an error recovery routine, to help guard against a failure that corrupts the microcontroller's instruction pointer which could otherwise cause random code to be executed.

Analog Issues in Digital Circuits

Digital circuits are made from analog components, and care has to be taken in design so that the analog nature of these underlying components doesn't dominate over the desired digital behavior. In particular, attention must be paid to all noise and timing margins, to parasitic inductances and capacitances, to proper filtering of power and ground connections, to electromagnetic coupling amongst datelines, and many other details. Inattention to these can cause intermittent problems such as "glitches", vanishingly-fast pulses that may trigger some logic but not others, "runt pulses" that do not reach valid switching (threshold) voltages, or unexpected ("uudecoded") combinations of logic states.

A corollary of the fact that digital circuits are made from analog components is the fact that digital circuits are slower to perform calculations than analog circuits that occupy a similar amount of physical space and consume the same amount of power. However, the digital circuit will perform the calculation with much better repeatability, due to the high noise immunity of digital circuitry.

Construction

A digital circuit is often constructed from small electronic circuits called logic gates. Each logic gate represents a function of Boolean logic. A logic gate is an arrangement of electrically controlled switches. The output is an electrical flow or voltage that can, in turn, control more logic gates. Logic gates often use the fewest number of transistors in order to reduce their size, power consumption and cost, and increase their reliability. Manufactured as integrated circuits, they are the least expensive implementation when made in large volumes. They are usually designed by engineers using electronic design automation software.

Another form of digital circuit is constructed from lookup tables, (many sold as "programmable logic devices", though other kinds of PLDs exist). Lookup tables can perform all the same functions as machines based on logic gates, but lookup tables can be easily reprogrammed without changing the wiring. This means that a designer can often repair errors without changing the arrangement of wires. Therefore, in small volume products, programmable logic devices are often the preferred solution. They are usually designed by engineers using electronic design automation software.

When the volumes are medium to large, and the logic can be slow, or involves complex algorithms or sequences, often a small microcontroller is programmed to make an embedded system. These are usually programmed by software engineers. When only one digital circuit is needed, and its design is totally customized, as for a factory production line controller, the conventional solution is a programmable logic controller, or PLC. These are usually programmed by electricians, using ladder logic.

Structure of Digital Systems

Engineers use many methods to minimize logic functions, in order to reduce the complexity, and thus the number of errors and the expense of digital circuits. The most widely used method is the application of a minimization algorithm within a CAD system, although historically, truth tables, Karnaugh Maps, and Boolean algebra have also been used.

Representations are crucial to an engineer's design of digital circuits. Some analysis methods only work with particular representations. The classical way to represent a digital circuit is with an equivalent set of logic gates. Another way, often with the least electronics, is to construct an equivalent system of electronic switches (usually transistors). One of the easiest ways is to simply have a memory containing a truth table. The inputs are fed into the address of the memory, and the data outputs of the memory become the outputs.

For automated analysis, these representations have digital file formats that can be processed by computer programs. Most digital engineers are very careful to select computer programs ("tools") with compatible file formats. To choose representations, engineers consider types of digital systems. Most digital systems divide into "combinatorial systems" and "sequential systems". A combinatorial system always presents the same output when given the same inputs. It is basically a representation of a set of logic functions.

A sequential system is a combinatorial system with some of the outputs fed back as inputs. This makes the digital machine perform a "sequence" of operations. The simplest sequential system is probably a flip flop, a mechanism that represents a binary digit or "bit". Sequential systems are often designed as state machines. In this way, engineers can design a system's gross behavior, and even test it in a simulation, without considering all the details of the logic functions.

Sequential systems divide into two further subcategories. "Synchronous" sequential systems change state all at once, when a "clock" signals changes state. "Asynchronous" sequential systems propagate changes whenever inputs change. Synchronous sequential systems are made of well-characterized asynchronous circuits such as flip-flops, that change only when the clock changes, and which have carefully designed timing margins.

The usual way to implement a synchronous sequential state machine is divide it into a piece of combinatorial logic and a set of flip flops called a "state register." Each time a clock signal ticks, the state register captures the feedback generated from the previous state of the combinatorial logic, and feeds it back as an unchanging input to the combinatorial part of the state machine. The fastest rate of the clock is set by the most time-consuming logic calculation in the combinatorial logic.

The state register is just a representation of a binary number. If the states in the state machine are numbered (easy to arrange), the logic function is just some logic that produces the number of the next state. In comparison, asynchronous systems are very hard to design because all possible states, in all possible timings must be considered. The usual method is to construct a table of the minimum and maximum time that each such state can exist, and then adjust the circuit to minimize the number of such states, and force the circuit to periodically wait for all of its parts to enter a compatible state. (This is called "self-resynchronization.") Without such careful design, it is easy to accidentally produce asynchronous logic that is “unstable”, that is, real electronics will have unpredictable results because of the cumulative delays caused by small variations in the values of the electronic components. Certain circuits (such as the synchronizer flip-flops, switch debouncers, and the like which allow external unsynchronized signals to enter synchronous logic circuits) are inherently asynchronous in their design and must be analyzed as such.

Almost all digital machines are synchronous designs because it is much easier to create and verify a synchronous design. However, asynchronous logic is thought to be superior, if it can be made to work, because its speed is not constrained by an arbitrary clock; instead, it simply runs at the maximum speed permitted by the propagation rates of the logic gates from which it is constructed. Building an asynchronous circuit using faster parts implicitly makes the circuit “go” faster. More generally, many digital systems are data flow machines. These are usually designed using synchronous register transfer logic, using hardware description languages such as VHDL or Verilog.

In register transfer logic, binary numbers are stored in groups of flip flops called registers. The outputs of each register are a bundle of wires called a “bus” that carries that number to other calculations. A calculation is simply a piece of combinatorial logic. Each calculation also has an output bus, and these may be connected to the inputs of several registers. Sometimes a register will have a multiplexer on its input, so that it can store a number from any one of several buses. Alternatively, the outputs of several items may be connected to a bus through buffers that can turn off the output of all of the devices except one. A sequential state machine controls when each register new data from its input.

In the 1980s, some researchers discovered that almost all synchronous register-transfer machines could be converted to asynchronous designs by using first-in-first-out synchronization logic. In this scheme, the digital machine is characterized as a set of data flows. In each step of the flow, an asynchronous “synchronization circuit” determines when the outputs of that step are valid, and presents a signal that says, “grab the data” to the stages that use that stage’s inputs. It turns out that just a few relatively simple synchronization circuits are needed.

The most general-purpose register-transfer logic machine is a computer. This is basically an automatic binary abacus. The control unit of a computer is usually designed as a microprogram run by a microsequencer. A microprogram is much like a player-piano roll. Each table entry or “word” of the microprogram commands the state of every bit that controls the computer. The sequencer then counts, and the count address the memory or combinatorial logic machine that contains the microprogram. The bits from the microprogram control the arithmetic logic unit, memory and other parts of the computer, including the microsequencer itself. In this way, the complex task of designing the controls of a computer is reduced to a simpler task of programming a relatively independent collection of much simpler logic machines.

Computer architecture is a specialized engineering activity that tries to arrange the registers, calculation logic, buses and other parts of the computer in the best way for some purpose. Computer architects have applied large amounts of ingenuity to computer design to reduce the cost and increase the speed and immunity to programming errors of computers. An increasingly common goal is to reduce the power used in a battery-powered computer system, such as a cell-phone. Many computer architects serve an extended apprenticeship as microprogrammers. “Specialized computers” are usually a conventional computer with a special-purpose microprogram.

Automated Design Tools

To save costly engineering effort, much of the effort of designing large logic machines has been automated. The computer programs are called "electronic design automation tools" or just "EDA." Simple truth table-style descriptions of logic are often optimized with EDA that automatically produces reduced systems of logic gates or smaller lookup tables that still produce the desired outputs.

Most practical algorithms for optimizing large logic systems use algebraic manipulations or binary decision diagrams, and there are promising experiments with genetic algorithms and annealing optimizations. To automate costly engineering effort, some EDA can take state tables that describe state machines and automatically produce a truth table for the combinatorial part of a state machine. The state table is a piece of text that lists each state, and the conditions that can exit that state.

It is common for the truth tables of such computer-generated state-machines to be optimized with logic-minimization software. This is a simple example of how complex logic machines are broken into simpler parts. Often, real logic systems are designed as a series of sub-projects, which are combined using a "tool flow". The tool flow is usually a "script", a simplified computer language that can invoke the software design tools in the right order.

Tool flows for large logic systems such as microprocessors can be thousands of commands long, and combine the work of hundreds of engineers. Writing and debugging tool flows is an established engineering specialty in companies that produce complex logic machines. The tool flow usually terminates in a detailed computer file or set of files that describe how to physically construct the logic machine. Often it consists of instructions to draw the transistors and wires on an integrated circuit or a printed circuit board.

Parts of tool flows are "debugged" by testing the outputs of simulated logic machines against expected inputs. The test tools take computer files with sets of inputs and outputs, and highlight discrepancies between the simulated behavior and the expected behavior. These test data are usually called "test vectors." Often, the test vectors are preserved and used in the factory to test that newly constructed logic machines work correctly.

Design for Testability

A large logic machine (say, with more than a hundred logical variables) can have an astronomical number of possible states. Obviously, in the factory, testing every state is impractical if testing each state takes a microsecond, and there are more states than the number of microseconds since the universe began. Unfortunately, this ridiculous-sounding case is typical. Fortunately, large logic machines are almost always designed as assemblies of smaller logic machines. To save time, the smaller sub-machines are isolated by permanently-installed "design for test" circuitry, and are tested independently.

One common test scheme known as "scan design" moves test bits serially (one after another) from external test equipment through one or more serial shift registers known as "scan chains". Serial scans have only one or two wires to carry the data, and minimize the physical size and expense of the infrequently-used test logic.

After all the test data bits are in place, the design is reconfigured to be in "normal mode" and one or more clock pulses are applied, to test for faults (e.g. stuck-at low or stuck-at high) and capture the test result into flip-flops and/or latches in the scan shift register(s). Finally, the result of the test is shifted out to the block boundary and compared against the predicted "good machine" result.

In a board-test environment, serial to parallel testing has been formalized with a standard called "JTAG" (named after the "Joint Testing Acting Group" that proposed it). Another common testing scheme provides a test mode that forces some part of the logic machine to enter a "test cycle." The test cycle usually exercises large independent parts of the machine.

Trade-offs

Several numbers determine the practicality of a system of digital logic. Engineers explored numerous electronic devices to get an ideal combination of speed, low cost and reliability. The cost of a logic gate is crucial. In the 1930s, the earliest digital logic systems were constructed from telephone relays because these were inexpensive and relatively reliable. After that, engineers always used the cheapest available electronic switches that could still fulfill the requirements.

The earliest integrated circuits were a happy accident. They were constructed not to save money, but to save weight, and permit the Apollo Guidance Computer to control an inertial guidance system for a spacecraft. The first integrated circuit logic gates cost nearly $50 (in 1960 dollars, when an engineer earned $10,000/year). To everyone's surprise, by the time the circuits were mass-produced, they had become the least-expensive method of constructing digital logic. Improvements in this technology have driven all subsequent improvements in cost.

With the rise of integrated circuits, reducing the absolute number of chips used represented another way to save costs. The goal of a designer is not just to make the simplest circuit, but to keep the component count down. Sometimes this results in slightly more complicated designs with respect to the underlying digital logic but nevertheless reduces the number of components, board size, and even power consumption.

For example, in some logic families, NAND gates are the simplest digital gate to build. All other logical operations can be implemented by NAND gates. If a circuit already required a single NAND gate, and a single chip normally carried four NAND gates, then the remaining gates could be used to implement other logical operations like logical and. This could eliminate the need for a separate chip containing those different types of gates.

The "reliability" of a logic gate describes its mean time between failures (MTBF). Digital machines often have millions of logic gates. Also, most digital machines are "optimized" to reduce their cost. The result is that often, the failure of a single logic gate will cause a digital machine to stop working.

Digital machines first became useful when the MTBF for a switch got above a few hundred hours. Even so, many of these machines had complex, well-rehearsed repair procedures, and would be nonfunctional for hours because a tube burned-out or a moth got stuck in a relay. Modern transistorized integrated circuit logic gates have MTBFs of nearly a trillion (1x10^12) hours, and need them because they have so many logic gates.

The "fan out" describes how many logic inputs can be controlled by a single logic output. The minimum practical fan out is about five. Modern electronic logic using CMOS transistors for switches have fan outs near fifty, and can sometimes go much higher.

The "switching speed" describes how many times per second an inverter (an electronic representation of a "logical not" function) can change from true to false and back. Faster logic can accomplish more operations in less time. Digital logic first became useful when switching speeds got above fifty hertz, because that was faster than a team of humans operating mechanical calculators. Modern electronic digital logic routinely switches at five gigahertz (5×10^9 hertz), and some laboratory systems switch at more than a terahertz (1×10^{12} hertz).

Non-electronic Logic

It is possible to construct non-electronic digital mechanisms. In principle, any technology capable of representing discrete states and representing logic operations could be used to build mechanical logic. Hydraulic, pneumatic and mechanical versions of logic gates exist and are used in situations where electricity cannot be used. The first two types are considered under the heading of fluidics. One application of fluidic logic is in military hardware that is likely to be exposed to a nuclear electromagnetic pulse (nuclear EMP, or NEMP) that would destroy any electrical circuits.

Mechanical logic is frequently used in inexpensive controllers, such as those in washing machines. Famously, the first computer design, by Charles Babbage, was designed to use mechanical logic. Mechanical logic might also be used in very small computers that could be built by nanotechnology.

Another example is that if two particular enzymes are required to prevent the construction of a particular protein, this is the equivalent of a biological "NAND" gate. The term digital signal is used to refer to more than one concept. It can refer to discrete-time signals that are digitized or to the waveform signals in a digital system.

Discrete-time Signals

Main Digital signals are digital representations of discrete-time signals, which are often derived from analog signals. An analog signal is a datum that changes over time—say, the temperature at a given location; the depth of a certain point in a pond; or the amplitude of the voltage at some node in a circuit—that can be represented as a mathematical function, with time as the free variable (abscissa) and the signal itself as the dependent variable (ordinate). A discrete-time signal is a sampled version of an analog signal: the value of the datum is noted at fixed intervals (for example, every microsecond) rather than continuously.

If individual time values of the discrete-time signal, instead of being measured precisely (which would require an infinite number of digits), are approximated to a certain precision—which, therefore, only requires a specific number of digits—then the resultant data stream is termed a digital signal. The process of approximating the precise value within a fixed number of digits, or bits, is called quantization.

A digital signal is a quantized discrete-time signal; a discrete-time signal is a sampled analog signal. In the Digital Revolution, the usage of digital signals has increased significantly. Many modern media devices, especially the ones that connect with computers use digital signals to represent signals that were traditionally represented as continuous-time signals; cell phones, music and video players, personal video recorders, and digital cameras are examples.

In most applications, digital signals are represented as binary numbers, so their precision of quantization is measured in bits. Suppose, for example, that we wish to measure a signal to two significant decimal digits. Since seven bits, or binary digits, can record 128 discrete values (viz., from 0 to 127), those seven bits are more than sufficient to express a range of one hundred values.

Waveforms in Digital Systems

In computer architecture and other digital systems, a waveform that switches between two voltage levels representing the two states of a Boolean value (0 and 1) is referred to as a digital signal, even though it is an analog voltage waveform, since it is interpreted in terms of only two levels. The clock signal is a special digital signal that is used to synchronize digital circuits. The image shown can be considered the waveform of a clock signal. Logic changes are triggered either by the rising edge or the falling edge.

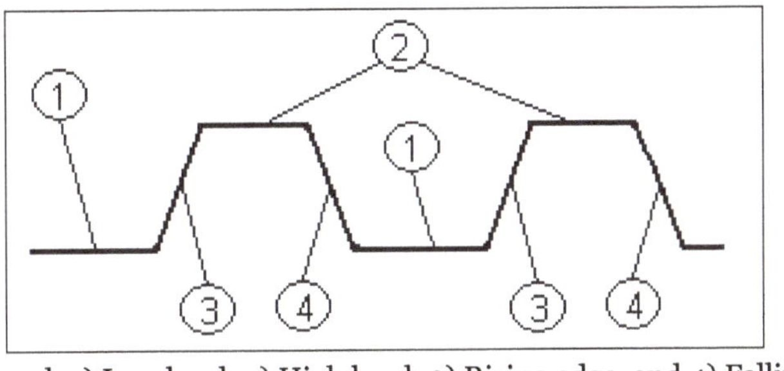

Digital signal: 1) Low level, 2) High level, 3) Rising edge, and 4) Falling edge.

Logic Voltage Levels

The two states of a wire are usually represented by some measurement of electric current: Voltage is the most common, but current is used in some logic families. A threshold is designed for each logic family. When below that threshold, the wire is "low," when above "high." Digital circuits establish a "no man's area" or "exclusion zone" that is wider than the tolerances of the components. The circuits avoid that area, in order to avoid indeterminate results.

Hobbyist frequency counter circuit built almost entirely of TTL logic chips.

It is usual to allow some tolerance in the voltage levels used; for example, 0 to 2 volts might represent logic 0, and 3 to 5 volts logic 1. A voltage of 2 to 3 volts would be invalid and would occur only in a fault condition or during a logic level transition, as most circuits are not purely resistive, and therefore cannot instantly change voltage levels. However, few logic circuits can detect such a fault, and most will just choose to interpret the signal randomly as either a 0 or a 1. The levels represent the binary integers or logic levels of 0 and 1. In active-high logic, "low" represents binary 0 and "high" represents binary 1. Active-low logic uses the reverse representation.

Digital Signal

A digital signal refers to an electrical signal that is converted into a pattern of bits. Unlike an analog signal, which is a continuous signal that contains time-varying quantities, a digital signal has a discrete value at each sampling point. The precision of the signal is determined by how many samples are recorded per unit of time. For example, the illustration below shows an analog pattern (represented as the curve) alongside a digital pattern (represented as the discrete lines). A digital signal is easily represented by a computer because each sample can be defined with a series of bits that are either in the state 1 (on) or 0 (off). Digital signals can be compressed and can include additional information for error correction.

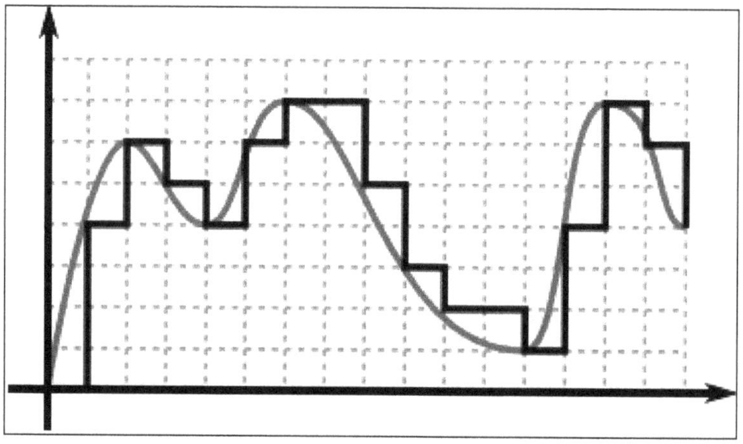

Logic Levels

Before examining digital signals, we must define logic levels. A logic level is a voltage level that represents a defined digital state.

- Logic High: The higher of two voltages, typically 5 volts.
- Logic Low: The lower of two voltages, typically 0 volts.

Logic Level	Voltage	True/False	On/Off	0/1
High	5 volts	True	On	1
Low	0 volts	False	Off	0

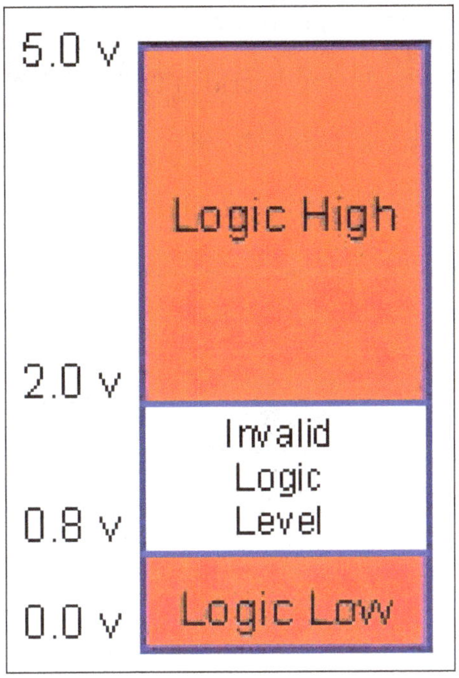

Example of Digital Signals

- Digital signal are commonly referred to as square waves or clock signals.
- Their minimum value must be 0 volts, and their maximum value must be 5 volts.
- They can be periodic (repeating) or non-periodic.
- The time the signal is high (t_H) can vary anywhere from 1% of the period to 99% of the period.

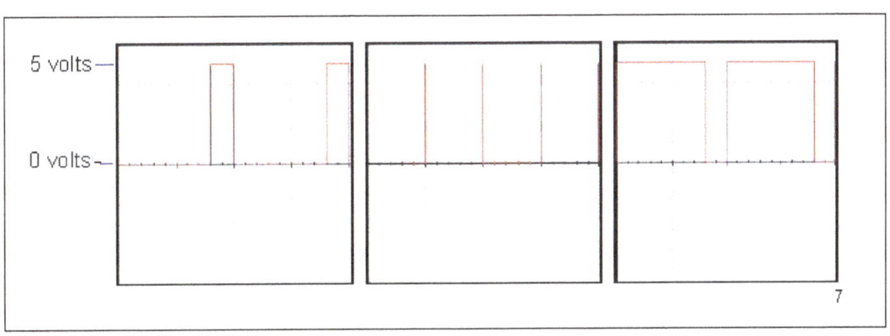

Parts of a Digital Signal

- Amplitude: For digital signals, this will always be 5 volts.
- Period: The time it takes for a periodic signal to repeat. (seconds)
- Frequency: A measure of the number of occurrences of the signal per second. (Hertz, Hz)
- Time High (t_H): The time the signal is at 5 v.

- Time Low (t_L): The time the signal is at 0 v.
- Duty Cycle: The ratio of t_H to the total period (T).
- Rising Edge: A 0-to-1 transition of the signal.
- Falling Edge: A 1-to-0 transition of the signal.

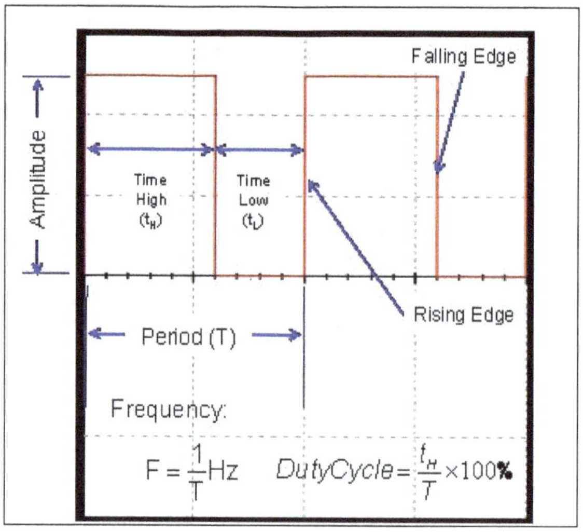

Advantages of Digital Signals

- Digital data can be easily compressed.
- Equipment that uses digital signals is more common and fewer expensive.
- These signals turn the moving instruments free from errors.
- You can edit the sound without altering the first copy.
- Digital signals can convey information with less noise, distortion, and interference.
- Digital signals can be reproduced easily in mass quantities at comparatively low costs.
- As the DSP operations are often altered that's why digital signal processing is more flexible.
- Digital signal processing is safer because digital information is often easily encrypted and compressed.
- Digital systems are more accurate, and therefore the probability of error occurrence is often reduced by employing error detection and correction codes.
- Digital signals are often easily stored on any magnetic media or optical media using semiconductor chips.
- Digital signals can be transmitted over long distances.

Disadvantages of Digital Signals

- Sampling may cause loss of information.

- A/D and D/A demands mixed-signal hardware.
- Processor speed is limited.
- Develop quantization and round-off errors.
- Systems and processing is more complex.
- A higher bandwidth is required for data communication in comparison to analog transmission of equivalent information.
- Digital systems and processing are typically more complex.

Arithmetic Circuits

Nowadays, computer is the most important part of our life. It consists of three following important units: (i) Input/Output devices, (ii) Memory and (iii) CPU. CPU is the heart of computer. It is central processing unit. Its basic element is ALU i.e. Arithmetic Logic Unit. It performs the arithmetic operations like addition, subtraction, multiplication, division etc. The logic operations like AND, OR, NOT etc. are performed. But addition and subtraction are basic operations. These are performed by arithmetic section of ALU. We are very familiar with decimal addition but computer does not do decimal addition. It converts the decimal number into binary numbers and then adds them. The other operations are multiple additions or subtraction. Subtraction is also an addition. Therefore addition is the most important operation in arithmetic circuits.

Half Adder

The most basic arithmetic operation is the addition of two binary digits. This simple addition consists of four possible elementary operations:

- 0 + 0 = 0
- 0 + 1 = 1
- 1 + 0 = 1
- 1 + 1 = 10

A combinational circuit that performs the addition of two bits is called a half adder. One that performs the addition of three bits (two significant bits and a previous carry) is a full adder. The names of the circuits stem from the fact that two half adders can be employed to implement a full adder. Let us consider x and y as the inputs to the half adder whereas Sum and Carry are the outputs. The Truth table, K-maps for sum and carry along with the logic diagram are shown in figure.

Table: Truth table for half adder.

x	y	Sum	Carry
0	0	0	0

0	1	1	0
1	0	1	0
1	1	0	1

	x'	x
y'		1
y	1	

K-map for sum.

	x'	x
y'		
y		1

K-map for carry.

Boolean expression for Sum: $Sum = x'y + xy'$

Boolean expression for Carry: $Carry = x.y$

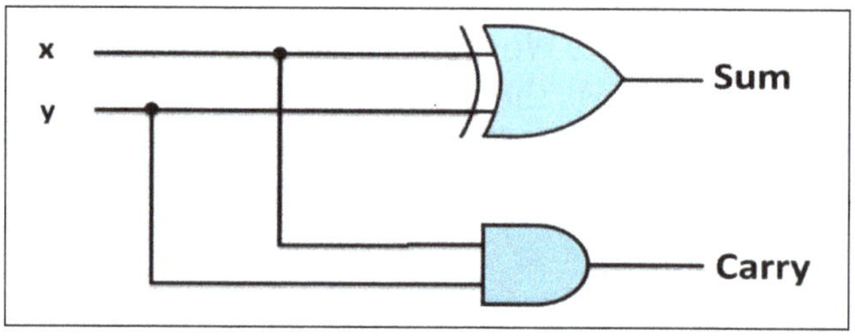

Truth table, K-maps for sum and carry along with the logic circuit for half adder.

Let us consider the truth table of half adder. There are four possible combinations for the inputs. When both x and y are at logic '0' then sum as well as carry are at logic '0'. If any one of the input either x or y is '1' the sum is at '1' and carry is '0'. When both the inputs are '1' then the sum is '0' and carry becomes '1'.

The k-map for sum and carry need to be prepared separately. The k-map is of 2-varibles for sum and carry as shown in figure. After plotting the outputs into the k-map the two Boolean expressions for sum and carry are obtained. These equations can be used to draw the logic diagram for half adder. The equation for sum indicate, there is a need of an EXOR gate and AND gate for the carry.

Full Adder

Full adder is a one of the most important blocks of an ALU. One that performs the addition of three bits (two significant bits and a previous carry) is a full adder. The names of the circuits stem from the fact that two half adders can be employed to implement a full adder. Let us begin with the preparation of Truth table for Full adder.

Table: Truth table for Full adder.

x	y	z	Carry	Sum
0	0	0	0	0
0	0	1	0	1
0	1	0	0	1
0	1	1	1	0
1	0	0	0	1
1	0	1	1	0
1	1	0	1	0
1	1	1	1	1
The third input Z=Carry in.				

	y'z'	y'z	yz	yz'
x'		1		1
x	1		1	

K-map for Sum.

	y'z'	y'z	yz	yz'
x'			1	
x		1	1	1

K-map for Carry.

Boolean expressions for Sum:

$$\begin{aligned} Sum &= z\oplus(x\oplus y) \\ &= z'(xy'+x'y)+z(xy'+x'y)' \\ &= z'(xy'+x'y)+z(xy+x'y') \\ &= xy'z'+x'yz'+xyz+x'y'z \end{aligned}$$

Boolean expressions for Carry:

$$\begin{aligned} Carry &= xy+xz+yz \\ &= xy+xy'z+x'yz \\ &= xy+z\,(x\oplus y) \end{aligned}$$

Let us consider the truth table of full adder: There are eight possible combinations for the inputs. When all the inputs are at logic '0' then sum as well as carry are at logic '0'. If any one of the input either x or y or z is '1' the sum is at '1' and carry is '0'. When any two of the inputs are '1' then the sum is '0' and carry becomes '1'. When all the inputs are '1' then the sum as well as carry is at logic '1'.

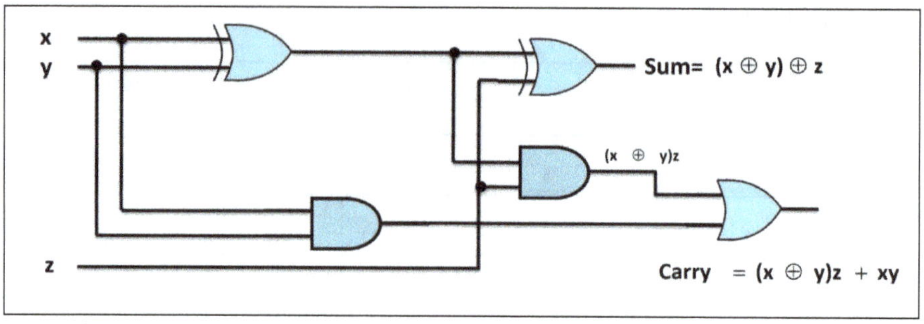

Truth table, K-maps for sum and carry along with the logic circuit for full adder.

The k-map for sum and carry need to be prepared separately. The k-map is of 3-varibles for sum and carry as shown in figure. After filling the outputs into the k-map and simplification, the two Boolean expressions for Sum and Carry are obtained. These equations can be used to draw the logic diagram for full adder. The full adders using two half adders and OR gate is obtained by expressions of sum and carry. In short, we require two EXOR gates, 2 AND gates and one OR gate to obtain the logic diagram for Full Adder.

Binary Adder

Multiple bits binary adder is obtained by using number of full adders connected in cascade. For the first full adder the least significant bit of Augend and Addend along with the carry in is the inputs. Its outputs are sum and carry out. The carry out is used as carry in for the next full adder. Other inputs for the next full adder are the next significant bits. It again generates the sum and carry. The process is repeated till the last MSB. Figure indicates the logic diagram for 4-bit binary adder.

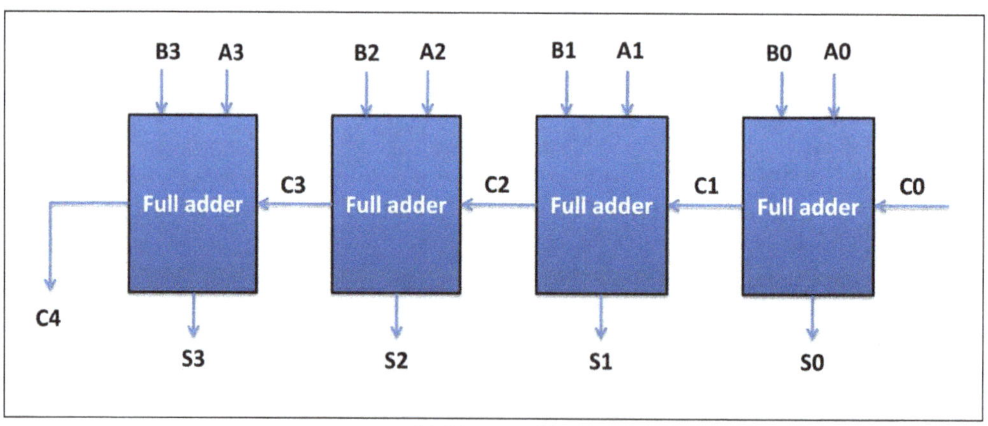

4 bit binary adder.

Let us consider the example of 4-bit binary addition. Let us take the decimal numbers 11 and 3 for the addition.

Augend 11

Addend 3

Sum 14

The binary addition for these numbers is shown below:

Augend (Ai)	1 0 1 1
Addend (Bi)	0 0 1 1
Input carry (Ci)	0 1 1 0
Sum (Si)	1 1 1 0
Output carry (Ci+1)	0

The result 1110 in binary is equivalent to decimal 14. So the logic circuit performs the addition of two numbers in binary.

Binary Subtractor

Let us now begin with process of subtraction. This simple subtraction consists of four possible elementary operations:

- 0 - 0 = 0
- 0 - 1 = 1 with borrow 1
- 1 - 0 = 1
- 1 - 1 = 0

A combinational circuit that performs the subtraction of two bits is called a half subtractor. One that performs the subtraction of three bits (two significant bits and a previous borrow) is a full subtractor. The names of the circuits stem from the fact that two half subtractors can be employed to implement a full subtractor.

Half Subtractor

Let us consider the truth table of half subtractor. There are four possible combinations for the inputs. When both x and y are at logic '0' then difference as well as borrow are at logic '0'. If A=0 and B=1 then difference is '1' and borrow is also at '1'. When A=1 and B=0 then the difference is 1 and borrow is 0. When both the inputs are '1' then the both difference and borrow are at logic '0'. From the truth table it is clear that, we need additional NOT gate while implementing the borrow output. There is a need of an EXOR gate and AND gate for implementation.

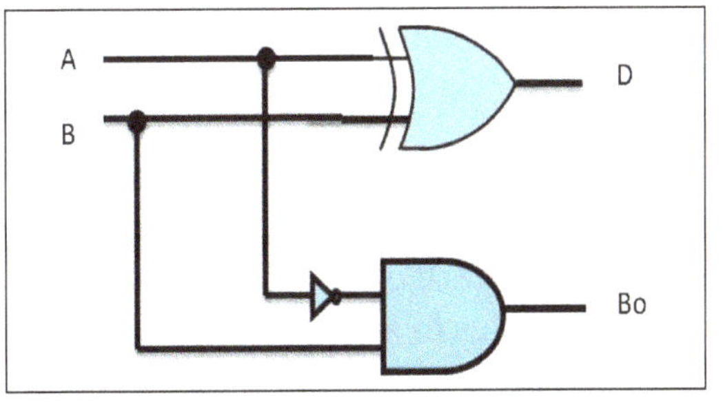

Table: Logic diagram and truth table of half Subtractor.

A	B	Difference (D)	Borrow (Bo)
0	0	0	0
0	1	1	1
1	0	1	0
1	1	0	0

Full Subtractor

One that performs the subtraction of three bits (two significant bits and a previous borrow) is a full subtractor. The names of the circuits stem from the fact that two half subtractors can be employed to implement a full subtractor.

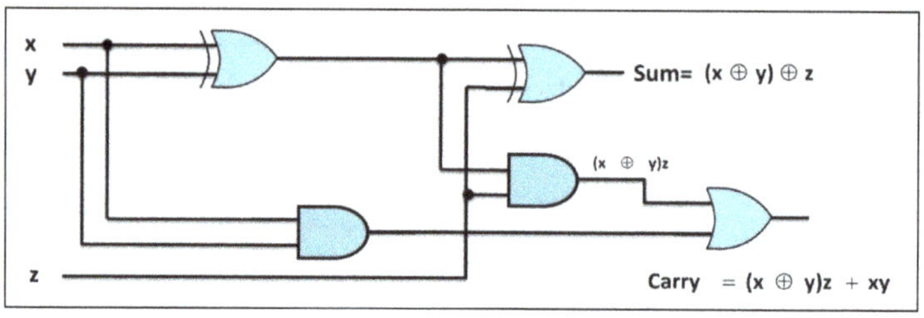

Universal Adder/Subtractor

There are two ways to perform binary subtraction. One way is to use the full subtractor and the other way is to implement 2's complement addition with discarding the carry (if generated). Normally, in ALU the subtraction is obtained by using 2's complement addition. 2's complement of the binary the number is obtained by first finding the 1's complement and adding binary '1' to it. For example:

	Decimal	Binary	2's complement addition
Minuend	11	1011	1011
Subtrahend	- 3	- 0011	+ 1101
------------------------------------		------------------	------------------
Difference	8	1000	1 1000

In 2's complement addition we need to convert the subtrahend in 2's complement form before addition. It is obtained by first obtaining the 1's complement and then addition binary '1' to it. In this example, subtrahend is 3. Its binary is 0011. Its 1'c complement is 1100 and 2's complement is 1101 (obtained by adding 1100 +1). Figure indicates the universal adder/subtractor circuit. It acts as adder when M=0 and provides subtraction when M=1.

Let us understand the working of this universal adder/subtractor. If control input M='0', all the XOR gates receives logic '0' as one of the inputs. This acts as an enable input and the gate transfers the data on other input as it is at the output. If control input M='1', all the XOR gates receives logic '1' as one of the inputs. This makes the gates to act like inverters. This will generate 1's complement of the input. Further the M'1' is also added, as it is given as carry input to the LSB adder.

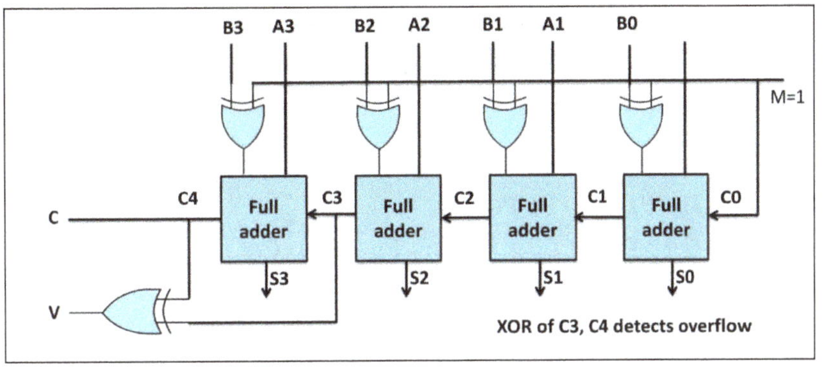

Carry Look-ahead Generator

The parallel adder accepts the data to be added at the same instant but the carries are not generated instantly. It takes time for the carry to propagate from LSB Full adder to MSB Full adder. The total propagation time is equal to propagation delay of a typical gate multiplied by the number of gates in the circuit. Inputs A3 and B3 are available as soon as input signals are applied to the adder. However, input carry C3 does not settle to its final value until C2 is available from the previous stage. Similarly, C2 has to wait for C1 and soon down to Co. Thus, only after the carry propagates and ripples through all stages will the last output S3 and carry C4 settle to their final correct value. This is called as carry propagation.

Carry Look-ahead Principle

The carry propagation time is the limiting factor on the speed of addition. The adder will always have some value at its output terminals. The correct value will be provided only when the signal propagates from inputs to outputs. Most of the arithmetic operations (subtraction, multiplication, division) are based on addition. The time consumed during addition is very critical. One solution for reducing the carry propagation delay time is to employ faster gates with reduced delays. Another solution is to increase the complexity of the circuit in such a way that the carry delay time is reduced. There are several techniques for reducing the carry propagation time in a parallel adder. The most widely used technique employs the principle of carry look ahead logic. The technique is used with the help of Full Adder obtained using Ex-OR, AND & OR gate.

If we define two new binary variables Gi – Carry generator & Pi – Carry propagator,

- Pi = Ai XOR Bi
- Gi = Ai Bi

The output sum and carry can respectively be expressed as,

- Si =Pi XOR Ci
- Ci+1 = Gi + PiCi

Gi called a carry generate. And it produces a carry of 1 when both Ai and Bi are 1. Regardless of the input carry Ci. Pi is called carry propagate, because it determines whether a carry into stage i will propagate into stage i+1. We write the Boolean equations for the carry outputs of each stage,

$$C_0 = \text{input carry}$$
$$C_1 = G_0 + P_0 C_0$$
$$C_2 = G_1 + P_1 C_1 = G_1 + P_1 G_0 + P_1 P_0 C_0$$
$$C_3 = G_2 + P_2 C_2 = G_2 + P_2 G_1 + P_2 P_1 G_0 + P_2 P_1 P_0 C_0$$

and,

$$S_0 = P_0 \text{ XOR } C_0$$
$$S_1 = P_1 \text{ XOR } C_1$$
$$S_2 = P_2 \text{ XOR } C_2$$
$$S_3 = P_3 \text{ XOR } C_3$$

Since the Boolean function for each output carry is expressed in sum-of-products form, each function can be implemented with one level of AND gates followed by an OR gate. All three Boolean functions for C1, C2 and C3 are implemented in the carry look ahead as shown in figure.

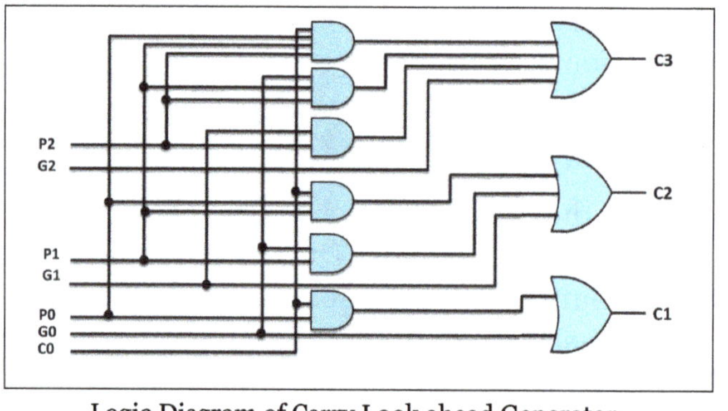

Logic Diagram of Carry Look ahead Generator.

Each sum output requires two XOR gates. The carries are propagated through the carry look ahead generator and applied as inputs to the second XOR gates as shown in figure. All output carries are generated after a delay through two levels of gates. Thus, outputs S1 through S3 have equal propagation delay times. Hence, the speed of addition is more.

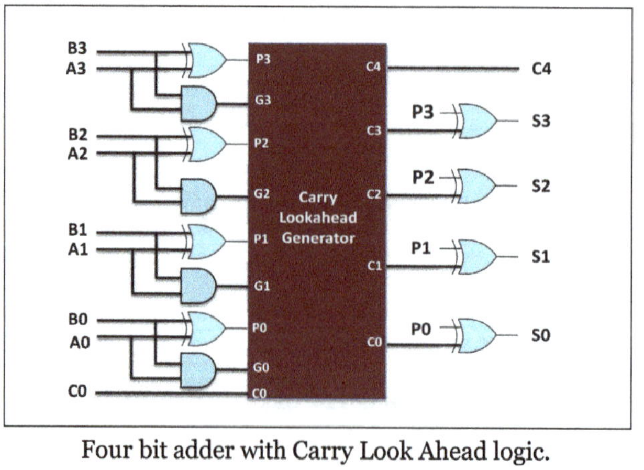

Four bit adder with Carry Look Ahead logic.

Digital Integrated Circuits

Digital integrated circuits do represent or stand for the practical sides of digital electronics. Digital electronic circuits hold/handle strategic or discrete signals such as 0 and 1. These two different states (logic 0 and logic 1) are low (logic 0) and high (logic 1). Digital integrated circuits employ digital multiplexers, counters, logic gates, and flip flops, among others.

In designed to accept a specific absolute voltage value (e.g., "true or false" logic operation) manner of designing digital electronic circuits. For instance, digital electronic circuits use two states, known as a binary circuit. These binary circuits come with binary quantities of "off" and "on" that represent 1 (right) and 0 (false). Unlike other courses, digital electronic circuits are economical and easy to manufacture or design.

Digital integrated circuits work at only a few states or defined levels as opposed to a wide range of signals. Digital integrated devices find use in a vast number of electronic devices, modems, computer networks, frequency networks, and computers. In short, some of the standard fundamental blocks of digital integrated circuits are logic gates. Unlike other types of courses or channels, digital integrated circuits only operate at few defined levels of signal amplitudes.

Digital Integrated Circuit Series

There exist a 7400-series and 4000-series of integrated circuits. Apt examples of 7 series include the following:

- 7400
- 7402
- 7404
- 7408
- 7432
- 7488

Types of 4 series includes:

- 4001
- 4009
- 4011
- 4030
- 4071
- 4077
- 4081

Examples of Integrated Circuit Series

There are different families of digital integrated circuits. When it comes to digital electronic circuits, a family refers to devices constructed using a combination of several automatic gates. A family consists of its individual or stand-alone logic levels and supply voltage power. Of importance to note is that different series of digital integrated circuits have their distinct advantages and disadvantages. Also, within every family, some voltages range may either be low or high. Below is a list of digital integrated circuit series/families:

- Diode Logic: When it comes to Diode Logic, the implementation of the entire logic gets done with the use of diodes and resistors. In the DL or logic diode, a diode or the primary purpose is to perform "and" "or" operator.
- Resistor-Transistor Logic: Regarding resistor-transistor thesis (RTL), the implementation of the entire logic is achieved using resistors and transistors. Resistors' electronics are not expensive, like the rest, and are too simple to design. The only drawback is that RTLs draw a significant amount of power.
- Diode-Transistor Logic: When it comes to Diode-transistor logic (DTL), the implementation of the entire logic is through transistors and diodes. DTL has got a couple of advantages over other resistor-transistor logic and diode logic. For instance, its diodes can efficiently perform OR and AND operations with ease. Another benefit associated with DTL is the fact that its OR operation is achievable with the use of diodes as opposed to resistors.
- Transistor-Transistor Logic: TTL or transistor-transistor logic has got its logic gates made around transistors. The transistor-transistor thesis employs the use of bipolar transistors and comes with different versions. The versions include Schottky TTL, High-speed TTL, Low power TTL, and Standard TTL. Transistor-transistor logic is one of the fastest available bi-polar circuits available currently.
- Emitter Coupled Logic: Here, the transistors do not go into deep saturation, meaning that there are no storage delays. Transistor-transistor logic finds use in applications that concern high-speed movement.

- Complementary Metal Oxide Semiconductor Logic: Complementary Metal Oxide Semiconductor Logic (CMOS) are well known for their low consumption of power coupled with high fan-out. CMOS finds use in several applications and microprocessor technologies. CMOS is also one of the most reliable logic families.

How Digital Integrated Circuits Work

Digital integrated circuits are small electronic components that brought about dramatic or unforgettable technologies in the history of man. Also known as a semiconductor chip, the digital integrated circuit has significantly unleashed change compared to the industrial revolution. But how do digital integrated circuits work? Digital integrated circuits consist of a combination of transistors, microprocessors, and diodes. They all play their different roles, such as storing voltage, controlling the flow of current, and providing memory to the entire system. These components/ devices work together in a digital integrated circuit to undertake different operations for effective operation or functionality.

Making your digital circuit is not a hard or rather tricky process, you may imagine. The process/ steps of producing a digital integrated circuit begin with a substantially single crystal of silicon shaped as reliable pipes. These pipes are also known as wafers. You will need to mark the biscuits into several identical rectangular or square areas.

After that, you will have to create components on every chip through doping some areas of the surface. Doping is achieved using a variety of processes. One of the most common methods is known as sputtering. The process of sputtering involves the firing of ions of the doping material/content to the silicon wafer. There is another process known as vapor deposition. Vapor deposition consists of the introduction of doping material in the form of gas and after that, letting it condense for the impurity atoms to form a thin film. The creation of the film occurs at the silicon wafers' surface.

Digital Integrated Circuits Directly Associated with PCBA

A Printed Circuit Board (PCB) is a board that connects electronic components by use of pads, conductive traces, and other features. They can be double-sided, single-sided, or multi-layer. On the other hand (Printed Circuit Board Assembly) is board presented after the entire parts and components get soldered and rightfully installed. A PCBA consisting of its parts and components easily accomplishes the automatic function it's designed to achieve.

Put in short, Printed Circuit Board Assembly) is board presented after the entire parts and components get soldered and rightfully installed. Printed Circuit Board Assembly is a board that comes out after the printing and mounting of components. These components include capacitors, transformers, and resistors, to mention but a few. During the process of assembly, an empty PCB board gets populated or stuffed with electronic components with the sole purpose of creating a functioning/working Printed Circuit Board Assembly) is board presented after the entire parts and components get soldered and rightfully installed.

Printed circuit boards (PCBs) find use in almost all electronic devices. However, a majority of PCBs are directly associated with the assembly of digital integrated circuits. Digital integrated circuits interconnect devices/components such as resistors, transistors, inductors, and capacitors.

There are some common problems/issues that surround digital integrated circuits. First, the methodology used to design them is resistor-transistor logic (RTL), which dates back to the 1980s. This technology is past its prime. Secondly, a matter to do with power is another issue. A lot of people do not understand questions concerning power so well.

Digital integrated circuits have numerous connections. If all these connections have the same voltage, no potential damage may occur. However, in case one of the links has got a different voltage/energy from the rest, some damage may occur. It is advisable to be very cautious regarding questions to with voltage. However, you can protect/shield them by wrapping the pins together in aluminum foil or connecting them to become a single piece.

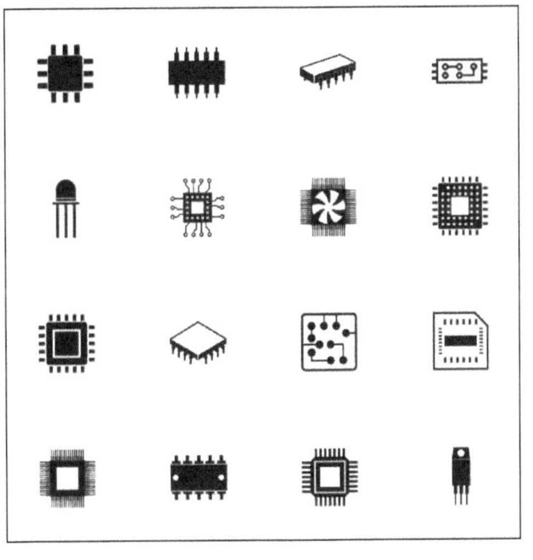

Combinational Circuits and Comparator

Combinational logic is a type of digital logic which is implemented using logic gates. The output of combinational circuit is the function of combination of present inputs only. Whereas in sequential circuit the output depends not only on the present input but also on the previous inputs i.e. history of inputs. In other words, a sequential circuit has memory while combinational circuit does not. Combinational circuit is able to generate an output simply from knowing what current input values are. Combinational circuits do not need to know the history of past inputs.

Let us take real life example of combinational circuit e.g. Car security system. In this example, the siren is turned on when the master switch is on and someone opens the door. If you close the door then the siren will turn off immediately. With this setup, the output, which is the siren, is depend only on the present inputs, which are the master and door switches.

A combinational circuit consists of input variables, logic gates, and output variables. The logic gates accept signals from the inputs and generate signals to the outputs. This process transforms binary information from the given input data to the required output data. Obviously, both input and output data are represented by binary signals, i.e. they exist in two possible values, one representing logic-1 and the other logic-0. A block diagram of a combinational circuit is shown in Figure.

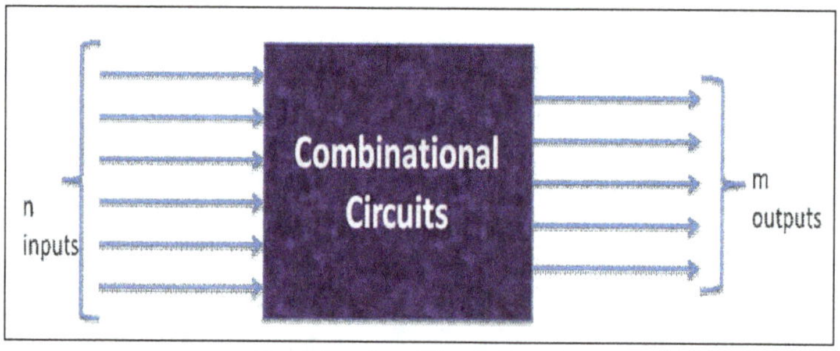

Combinational logic circuit.

Combination Logic Circuits are made up from basic gates (AND, OR, NOT) or universal gates (NAND, NOR) gates that are "combined" or connected together to produce more complicated logic circuits. These logic gates are the building blocks of combinational logic circuits. In these circuits "the outputs at any instant of time depends on the inputs present at that instant only." For n input variables, there are 2^n possible combinations of binary input values. For each possible input combination, there is one and only one possible output combination. A combinational circuit can be described by m Boolean functions, one for each output variable. Each output function is expressed in terms of the n input variables.

Types of Combinational Circuits

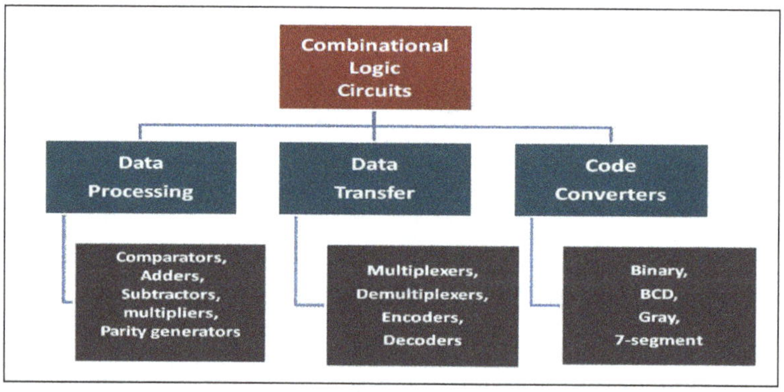

At any time a combinational logic circuit's outputs depends only upon the combination of its inputs at that time. It is important to note that output is not influenced by previous inputs. Figure shows the classification scheme of Combinational circuits. There are three major types of combinational circuits:

- Data processing circuits that process or transform data e.g. Adder, subtractor, comparator, multiplier etc.
- Data transfer circuits to control the flow of logic e.g. Multiplexer, demultiplexer, encoder, decoder etc.

- Code converters for conversion from binary to any other code and vice versa e.g. Binary, BCD, Gray, 7-segment converters.

Design Procedure of Combinational Circuits

For the design of Combinational digital circuits basic gates (AND, OR, NOT) or universal gates (NAND, NOR) are used. The design of combinational circuits begins with problem definition and ends with a logic circuit or set of Boolean expressions.

- State the problem clearly.
- Determine the available inputs and required outputs from problem.
- Assign a letter symbols to input and outputs of the circuits.
- Derive the truth table that relates inputs and outputs.
- Write the Boolean functions for the output and further simplify to get reduced number of variables.
- Draw the logic diagram.

Analysis Procedure of Combinational Circuits

In many applications, logic diagram is available but correlation between input and output is not known. The analysis of combinational circuits is the reverse process.

- Take a closer look at given combinational circuit.
- Identify the available inputs and outputs and label them appropriately.
- Find the output Boolean function for each gate.
- Obtain the output Boolean functions and/or truth table for the given circuits in terms of input variables.

Comparator

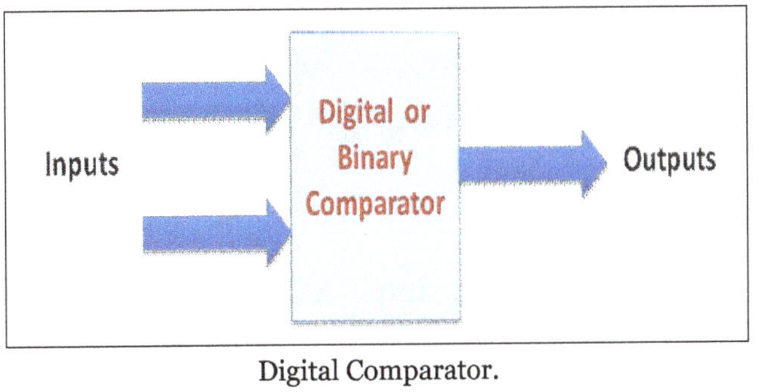

Digital Comparator.

Comparator is a combinational logic circuit that compares the magnitudes of two binary quantities to determine which one has the greater magnitude. In other word, a comparator determines the relationship of two binary quantities. An exclusive-OR gate can be used as a basic comparator.

Digital comparators are also called as binary or logic comparators. Comparator is a very useful combinational circuit capable of comparing two numbers as input in binary form and determines whether one number is greater than, less than or equal to other number. This logic circuit is used for testing whether the binary number at one input is greater than or less than or equal to another binary number. There are two types of comparators:

- Equality or Identity Comparator.
- Magnitude or inequality comparator.

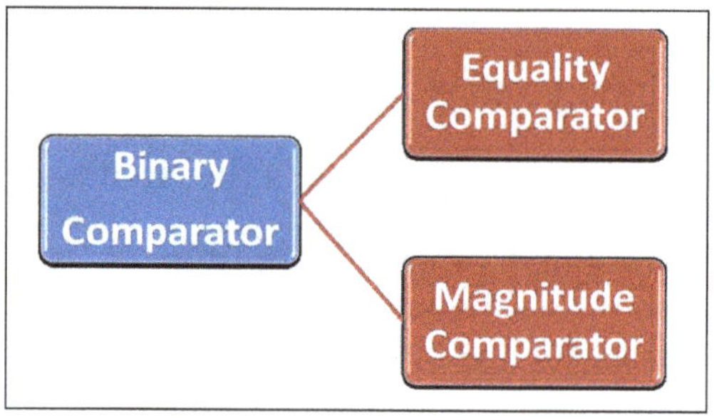

Equality Comparator

Equality comparator requires two input binary numbers A and B for comparison as shown in figure. There is only one output indicating whether the input numbers are equal or not. Output of the comparator is at logic '1' if the two input words A and B match otherwise the output remains at logic '0'. Out of the 2^n combinations, there is only one input combination that is correct and other combinations will result in inequality. This comparator can be used for simple circuits like electronic lock and security systems where binary password consisting of n bits compared with another preset code. Let us discuss the implementation of simple 1-bit equality comparator.

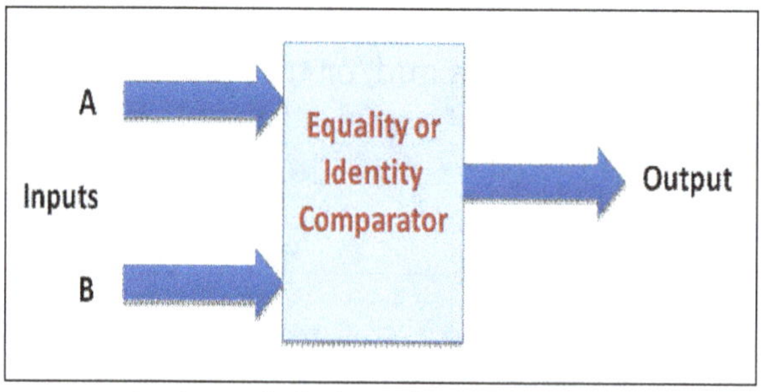

Logic symbol of equality comparator.

1-bit Equality Comparator

Let us study 1-bit equality comparator using simple XOR circuit. The truth table of XOR indicates the operation of Identity or Equality comparator. When both the inputs of the comparator are equal then the output is active LOW indicating matching of the input code, otherwise output is at logic '1' indicating the mismatch.

Let us study ACTIVE HIGH 1-bit equality comparator using simple XNOR circuit as shown in figure below. The truth table of XNOR indicates the operation of Identity or Equality comparator. If two inputs are identical then the output of comparator is '1' else provides '0' indicating mismatch in the input code.

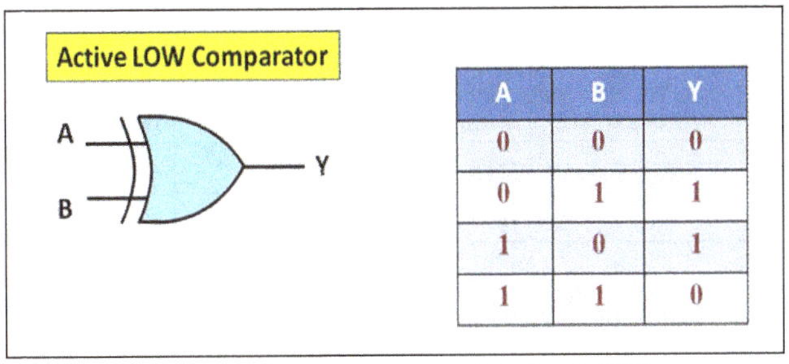

A	B	Y
0	0	0
0	1	1
1	0	1
1	1	0

1 bit Equality comparator (Active LOW).

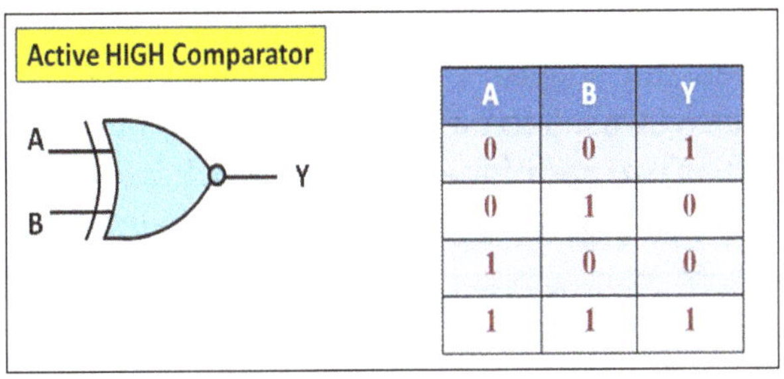

A	B	Y
0	0	1
0	1	0
1	0	0
1	1	1

1 bit Equality comparator (Active HIGH).

2-bit Equality Comparator

In two bit equality comparator as shown in figure, two bit binary numbers A and B represented as A1A0 and B1B0 are compared. Here A1 and B1 are the most significant bits, and A0 and B0 are the least significant bits of numbers A and B respectively. A1 is compared with B1 and A0 is compared the B0. If A1 is equal to B1 and A0 is equal to B0 output of a two bit comparator will be HIGH. If A1 is not equal to B1 or A0 is not equal to B0 output will be LOW as shown in the table.

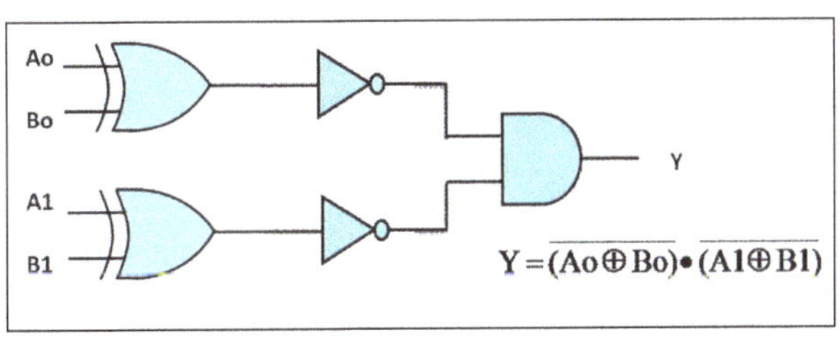

Inputs				Outputs
A1	A0	B1	B0	A=B
0	0	0	0	1
0	0	0	1	0
0	0	1	0	0
0	0	1	1	0
0	1	0	0	0
0	1	0	1	1
0	1	1	0	0

Class-B State Machine

Let us now talk about Moore machine. The class B state machine is known as Moore circuit, named after logic designer Moore. The basic property of a Moore circuit is that its output is strictly a function of present state (PS) of the sequential circuit. Careful observation of the Moore machine shows that the feed-forward mechanism is absent in the state machine. The output is,

$$Y = g(PS).$$

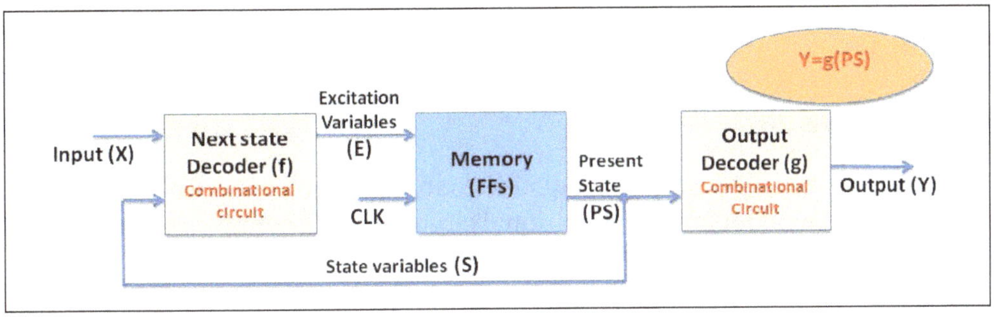

Class – B (Moore) machine.

Both Mealy and Moore machines are widely used. In some cases, it is possible to get circuits that are combination of both i.e. some outputs are dependent on both the inputs and present state of state machine and other outputs depend only on present state of the state machine.

Class-C State Machine

In the Class C state machine, the output decoder is not required to generate the output. The output of the state machine is nothing but the present state of the state machine. The generalized model for counter represents basically the class C state machine. In such circuit the events to be counted are entered through the next state decoder.

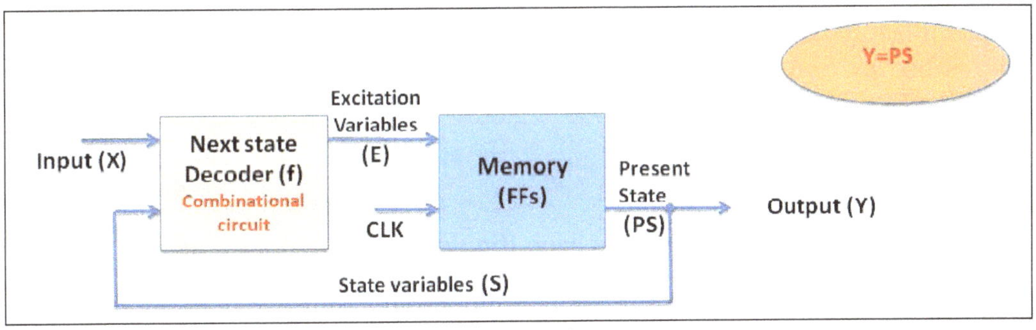

Class – C machine.

Class-D State Machine

In class D state machine, the outputs are directly obtained from the output of memory i.e. the present state of the state machine. The comparison of class D with class C state machines indicates that there is no feed-back of PS to decide the next state. The next state is decided directly by the input combination. Example of such circuit is the pre-settable counter where starting

0	1	1	1	0
1	0	0	0	0
1	0	0	1	0
1	0	1	0	1
1	0	1	1	0
1	1	0	0	0
1	1	0	1	0
1	1	1	0	0
1	1	1	1	1

Magnitude Comparator

Magnitude comparator as the name suggest compares only the magnitude of the two binary numbers and not the sign. It can also be used to indicate the equality of the input numbers. In addition, the magnitude comparators has two additional outputs to indicate whether input A is greater that input B and whether input A is less than input B as shown in figure.

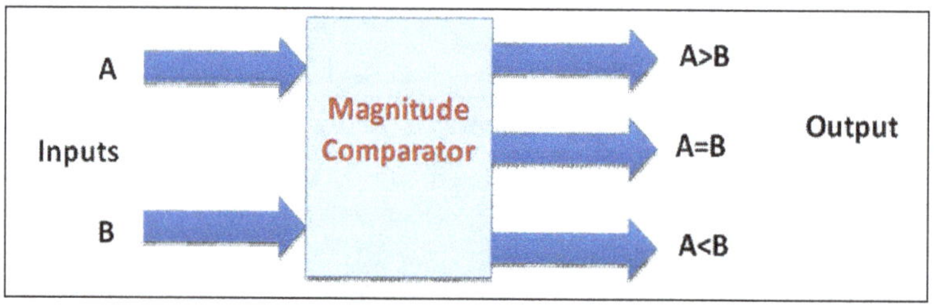

1-bit Magnitude Comparator

In 1-bit magnitude comparator, A, B are two 1 bit number and there are three outputs corresponding to A>B, A=B and A<B respectively as shown in figure. The Boolean expressions for 1-bit magnitude comparator are,

$$C = A \bullet \overline{B}$$

$$D = \overline{A \oplus B}$$

$$E = \overline{A} \bullet B$$

For implementation of this comparator, XNOR gate, AND gates and inverters are also used. It is very easy to derive the Boolean expressions from the truth table by visual inspection.

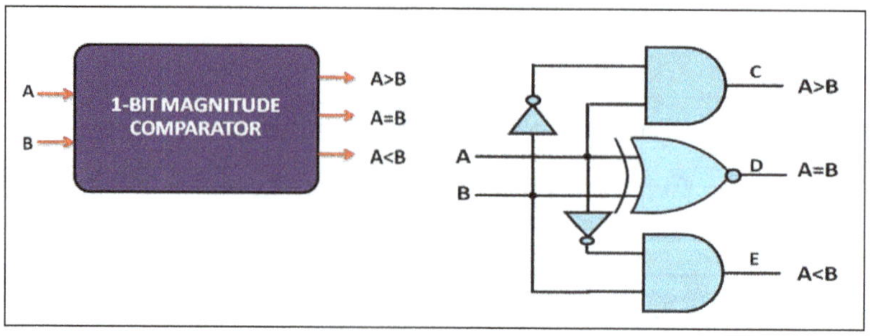

Table: 1-bit magnitude comparator.

Inputs		Outputs		
A	B	A>B	A=B	A<B
0	0	0	1	0
0	1	0	0	1
1	0	1	0	0
1	1	0	1	0

2-bit Magnitude Comparator

A 2-bit comparator compares two binary numbers, each of two bits and produces their relation such as one number is equal or greater than or less than the other. The figure below shows the block diagram of a two-bit comparator which has four inputs and three outputs.

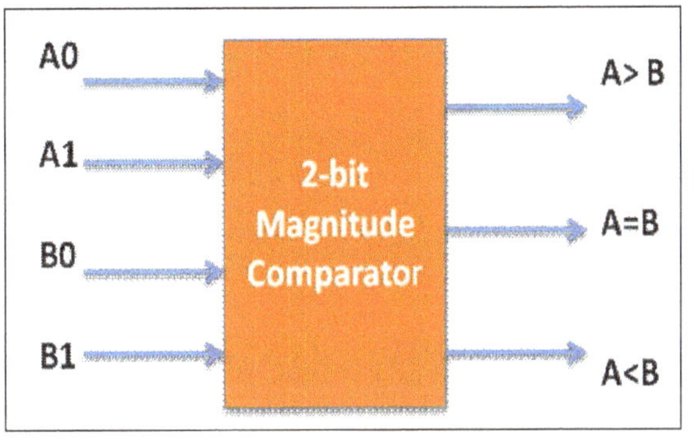

Table: 2-bit magnitude comparator.

Inputs				Outputs		
A1	A0	B1	BO	A>B	A=B	A<B
0	0	0	0	0	1	0
0	0	0	1	0	0	1
0	0	1	0	0	0	1
0	0	1	1	0	0	1
0	1	0	0	1	0	0
0	1	0	1	0	1	0
0	1	1	0	0	0	1
0	1	1	1	0	0	1
1	0	0	0	1	0	0
1	0	0	1	1	0	0
1	0	1	0	0	1	0
1	0	1	1	0	0	1
1	1	0	0	1	0	0
1	1	0	1	1	0	0
1	1	1	0	1	0	0
1	1	1	1	0	1	0

The figure shows the block diagram of a two-bit comparator which has four inputs and three outputs. The first number A is designated as A = A_1A_0 and the second number is designated as B = B_1B_0. This comparator produces three outputs as (A>B), (if A = B) and (A<B).

Let us prepare the K-map for each output so as to implement the 2-bit magnitude comparator. The K-map and Boolean expression for the outputs A>B, A=B and A>B of comparator are shown in figure.

	B1'B0'	B1'B0	B1B0	B1B0'
A1'A0'				
A1'A0	1			
A1A0	1	1		1
A1A0'	1	1		

A>B: = A1B1' + A1'A0B1'B0'+A1A0B1B0'
= A1B1' + A0B0' (A1'B1'+A1AB1)
= A1B1' + A0B0' Y1

	B1'B0'	B1'B0	B1B0	B1B0'
A1'A0'	1			
A1'A0		1		
A1A0			1	
A1A0'				1

A=B: = A1'A0'B1'B0'+A1'A0B1'B0+A1A0B1B0+A1A0'B1B0'
= (A1'B1'+A1B1) (A0'B0'+A0B0)
= Y1Y0

	B1'B0'	B1'B0	B1B0	B1B0'
A1'A0'		1	1	1
A1'A0			1	1
A1A0				
A1A0'			1	

A<B: = A1'B1+A1'A0'B1'B0+A1A0'B1B0
= A1'B1'+A0'B0 (A1'B1'+A1B1)
= A1'B1'+A0'B0 Y1

Obtaining Boolean expression for 2-bit magnitude comparator.

By using above obtained Boolean equation for each output, the logic diagram can be implemented by using four NOT gates, seven AND gates, two OR gates and two Ex-NOR gates. Figure shows the logic diagram of a 2-bit comparator using basic logic gates. It is also possible to construct this comparator by cascading of two 1-bit comparators.

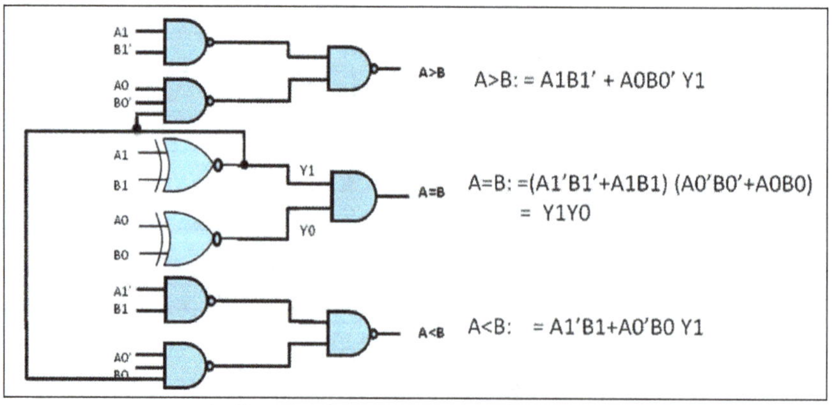

Logic diagram for 2-bit magnitude comparator.

A 4-bit Magnitude Comparator

Magnitude comparator can be used to compare two four-bit words. The two 4-bit numbers are A = A3 A2 A1 A0 and B= B3 B2 B1 B0 where A3 and B3 are the most significant bits. It compares each of these bits in one number with bits in that of other number and produces one of the following outputs as A = B, A < B and A>B. The output function table of 4-bit magnitude comparator is shown in Figure.

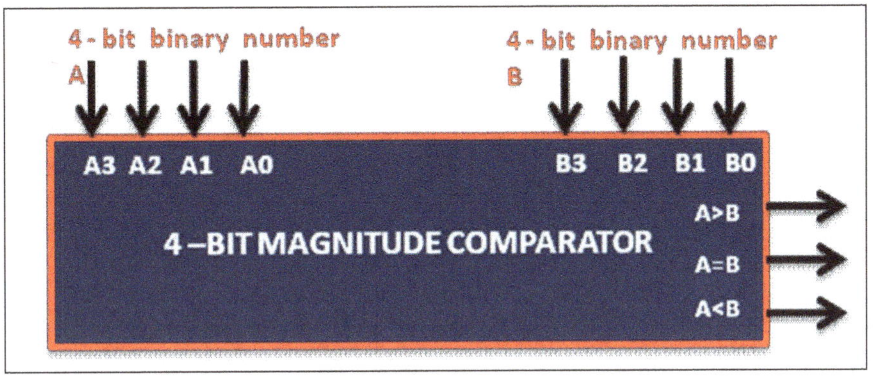

Logic symbol of 4-bit magnitude comparator.

Table: Function table of 4-bit magnitude comparator.

INPUT				OUTPUTS		
A3 B3	A2 B2	A1 B1	A0 B0	A > B	A = B	A < B
A3 > B3	X	X	X	1	0	0
A3 < B3	X	X	X	0	0	1
A3 = B3	A2 > B2	X	X	1	0	0
A3 = B3	A2 < B2	X	X	0	0	1
A3 = B3	A2 = B2	A1 > B1	X	1	0	0
A3 = B3	A2 = B2	A1 < B1	X	0	0	1
A3 = B3	A2 = B2	A1 = B1	A0 > B0	1	0	0
A3 = B3	A2 = B2	A1 = B1	A0 < B0	0	0	1
A3 = B3	A2 = B2	A1 = B1	A0 = B0	0	1	0

If A3 = 1 and B3 = 0, then A is greater than B (A>B), Or If A3 and B3 are equal, and if A2 = 1 and B2 = 0, then A > B. Or If A3 and B3 are equal & A2 and B2 are equal, and if A1 = 1, and B1 = 0, then A > B. Or If A3 and B3 are equal, A2 and B2 are equal and A1 and B1 are equal, and if A0 = 1 and B0 = 0, then A > B.

IC 7485

ICs are available in TTL as well as CMOS logic family ICs and some of these ICs include IC 7485 (4-bit comparator), IC 4585 (4-bit comparator in CMOS family) and IC 74AS885 (8-bit comparator). Figure indicates an IC for 4-bit magnitude comparator. The 74LS85 compares two unsigned 4-bit binary numbers the unsigned numbers are A_3, A_2, A_1, A_0 and B_3, B_2, B_1, B_0.

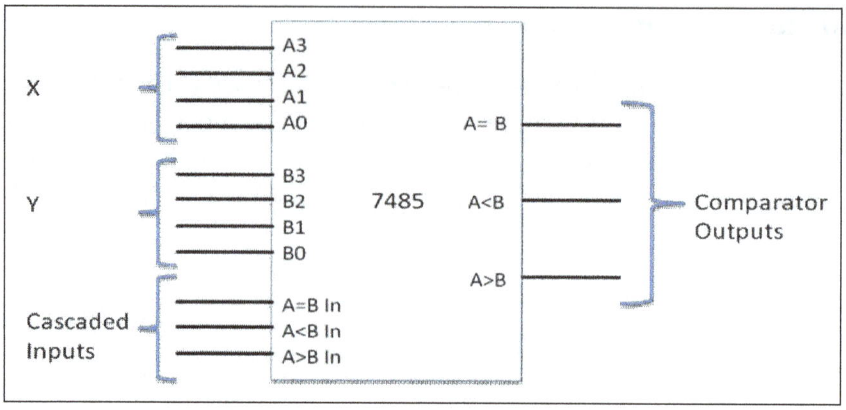

IC 7485- A 4 bit magnitude comparator.

Cascading Comparators

Cascading inputs provide a means for expanding the comparison operation to more than four bits by cascading two or more 4-bit comparators. When two comparators are to be cascaded, the outputs of the lower-order comparator are connected to corresponding inputs of the higher-order comparator. The comparator on the left is comparing lower 4 bits and right comparator compare higher order nibble.

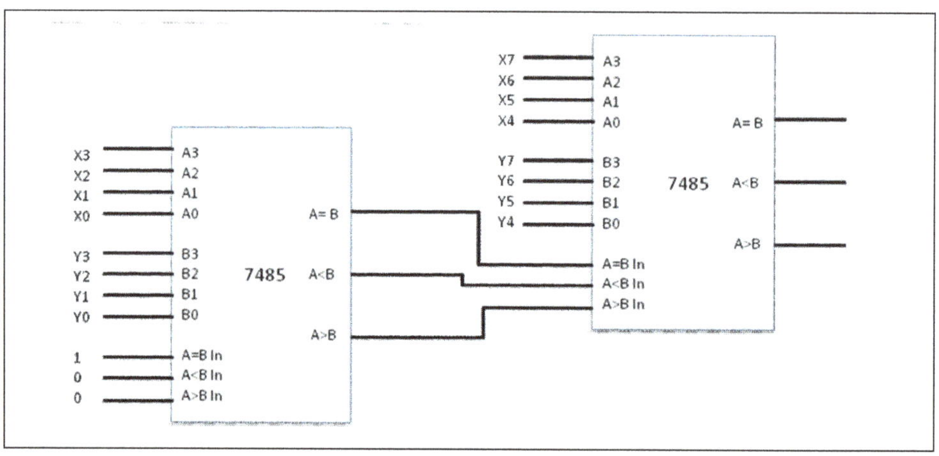

Cascading of Magnitude comparator.

Applications of Comparators

Comparators are used in control applications in which the binary numbers representing physical variables such as temperature, position, etc. are compared with a reference value. Then the outputs from the comparator are used to drive the actuators so as to make the physical variables closest to the set or reference value. In addition, comparators are also used in following applications:

- Arithmetic logic unit (ALU).
- Compare two binary words in microprocessors, computers for interfacing the external devices.
- Analog to digital converters (ADC).

- In Process Control Applications & for Servo-motor control.
- Password verification, biometric applications, for searching and surfing.

Sequential Circuits and Finite State Machine

The sequential circuit can be represented by a basic modeling structure so as to understand how flip flops play important role in sequential circuits. Models provide a means to represent an idea, concept or process mathematically or graphically. This helps in enhancing the understanding of sequential circuits. Figure shows how combinational circuits and memory (Flip flops) are correlated to each other to form sequential circuit.

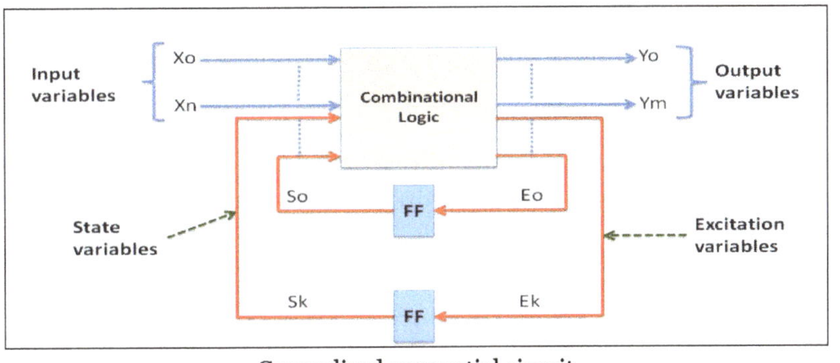

Generalized sequential circuit.

The universal model includes all the elements of sequential circuit. Combinational circuits accepts binary input variables Xo to Xn and generates output variables Yo to Yn. In addition, the system generates excitation variables Eo to Ek and accepts state variables So and Sk as past inputs (stored in memory i.e. flip flop). Thus, the output of sequential circuit is the function of present input conditions (Xi) and past inputs provided by state variables (Si). The memory elements will always appear in the feedback path.

To study a sequential circuit, it is important to know the internal state or condition of the circuit. Flip flops are used in feedback path with combinational circuits to form sequential circuit. This circuit follows as sequence of internal states. Each internal state is characterized by a set of allowable states. The input conditions determine which of these allowable successive states the system can undergo. The sequential circuits are classified into two categories: synchronous sequential and asynchronous sequential circuits. In synchronous sequential circuits, the contents of memory i.e. flip flops can be changed only at rising or falling edge of clock signal. In asynchronous sequential circuit the contents of memory can be changed at any instant of time.

Let us now talk about different terminologies used in state machine. These are called state machine notations. There are five different terminologies: Input variables, Output variables, State variables, excitation variables and the state.

- Input variable: All the variables which are external to the sequential circuit are known as input variables. There are finite number of input variables represented by the symbol X as, X = (Xo, X1,X2,.....XN).

- Output variable: All the variables that exit the sequential circuit are referred to as output variables. There is finite number of output variables represented by the symbol Y as, Y = (Yo, Y1, Y2,..... Yn).
- State variable: The output of memory (flip flops) defines the state of sequential machine. The output of flip flop defines the state variables. There are finite number of state variables represented as, S = (So, S1,...Sk).
- Excitation variable: Excitation variables are defined as the inputs to the memory i.e. flip flops. The name excitation is used because the variable excites the memory to change its state. When the flip flops are used as memory, the excitation variables are the inputs to the flip flops (JK, SR, D or T). State variables are a function of excitation variables. Excitation variables (E) are the function of input variables (X) and state variables (S).
- State: The state of sequential circuit is defined by the content of memory. The state is defined by Q outputs. Each state of sequential circuit should be unique and unambiguous. For k flip flops there will be 2k number of possible states.
- Present state and Next State: State machines normally use synchronous sequential circuit. Usually, rising or falling edge triggered flip flops are used. Flip flops can change state only on clock pulse edge hence it is necessary to distinguish state variables before and after the clock pulse. State variable X can be represented as Xt before the arrival of clock pulse and as Xt+1 after the arrival of clock pulse which is shown in the figure.

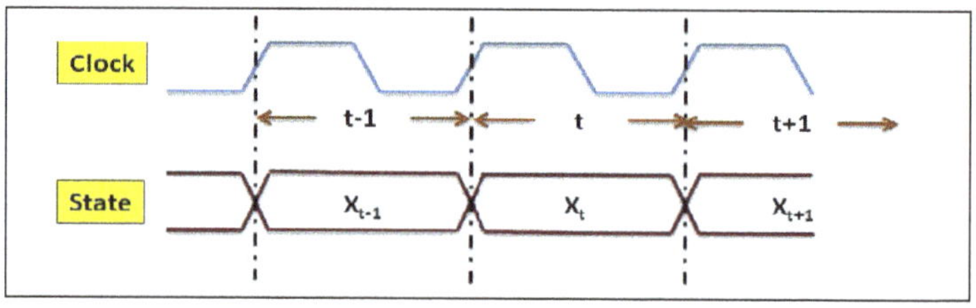

State variables (Present state and Next state) with Clock.

The status of all the state variables at any instant of time ‘t’, before the application of next clock edge is called a present state. The present state is taken as a reference point with respect to time. The status of all the state variables at some time ‘t+1’ after the application of clock edge is called as a next state. The next state is represented by flip flop status after a particular clock ‘t’. Present state and next states are normally used to prepare the state table of the state machine.

To study sequential circuit, it is necessary to know the internal state or condition of the circuit. Flip flops, counters and shift registers are all sequential logic circuits. All these circuits follow a sequence of internal states. The sequence of internal states is produced a sequence of inputs. Each of the input state is characterized by a set of allowable successive states. The input conditions determine which of these allowable successive states are generated next.

Let us now discuss the state diagram and state table for a sequential circuit. A state diagram is a graphical representation of the sequential circuit. Individual states are represented by a circle with an identifying symbol located inside. Let us consider a simple circuit having two stable states ‘0’ and ‘1’. This circuit has an external input X. The states of the circuit are shown by encircled numbers ‘0’ and’1’. Changes from one state to another are indicated by an arc with an arrow head at one end. Input condition that causes state changes are written adjacent to the directed arc. It is possible to represent state and self-transitions in the state diagram.

When the circuit is in state '0' and if input X = 0 then the circuit remains in the state '0'. Similarly if input X becomes 1, then circuit makes transition from state 0 to state 1. When the circuit is in state 1, if X becomes 1 then circuit remains in the same state then self-transition takes place. When X changes 0 then the circuit switches to state 0. This results in state transition for the circuit.

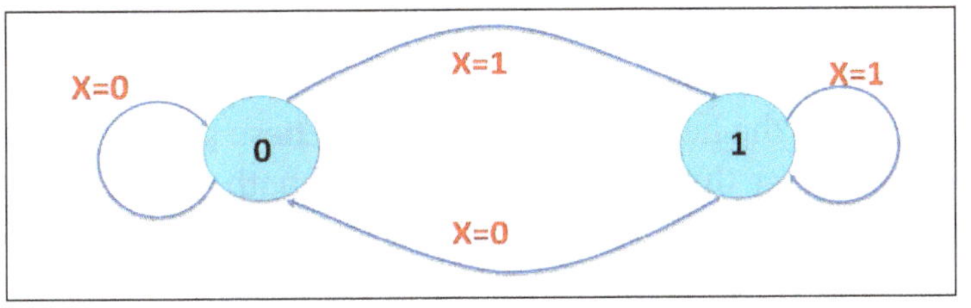

State diagram for a sequential circuit.

The behavior of the sequential circuit can also be represented by a state table. State tables are tabular forms of the state diagrams. First column in the state table represents all possible present states. There are a series of next state columns one for each input combination. The purpose of state table is to indicate the state transition. A two state, one input variable state diagram is converted into state table.

Present State (Qn)	Next State (Qn+1)	
	X=0	X=1
0	0	1
1	0	1

First column indicates two possible present states '0' and '1'. The circuit will change the state depending upon the value of input variable X. When the present state of the circuit is '0' and X is also '0', the next state of the circuit remains '0'. But is X='1', the next state of the circuit remains in state '0'. Similarly when present state is '1' and input variable x='1' then the next state of the circuit remains in state '1'. But if X=0 then next state is '0'. State tables give complete information about present states, next states, inputs and outputs of a sequential circuit.

Finite State Model

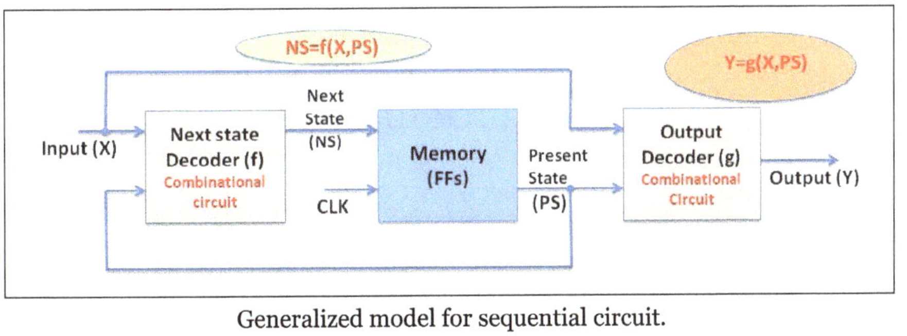

Generalized model for sequential circuit.

A sequential circuit has number of output states corresponding to different input and output combinations. For example, a decade counter produces 10 different states. Such a circuit is called as a finite state machine. Let us consider a structure of finite state machine as shown in this diagram. This is basically a generalized sequential circuit model. It has input-output combinational logic, popularly known as the next state decoder. It decodes the input (X) and the present state (PS) of the machine and generates the next state. The next state is a function of both input and the present state.

Next State = f(X,PS)

The next state information becomes the present state at the memory which stores it. As the generation of next state and present state is a continuous process, present state information is fed back to generate the next state with the help of present inputs. These present input conditions are also forwarded to the output decoder through a feed –forward path. These inputs with present state generate the necessary control signal at the output of the state machine. The output of FSM is given by,

$$Y = g(x, PS).$$

Finite state machines are classified according to the feed-back, feed-forward as well as other connections as:

- Class-A (Mealy) state machine,
- Class-B (Moore) state machine,
- Class-C state machine,
- Class-D state machine,
- Class-E state machine.

In the model of finite state machine, the next state (NS) is also referred to as excitation variables (E) for exciting or triggering the flip flops present in the memory. Similarly, the present state generates the state variables (S) for the next state decoder.

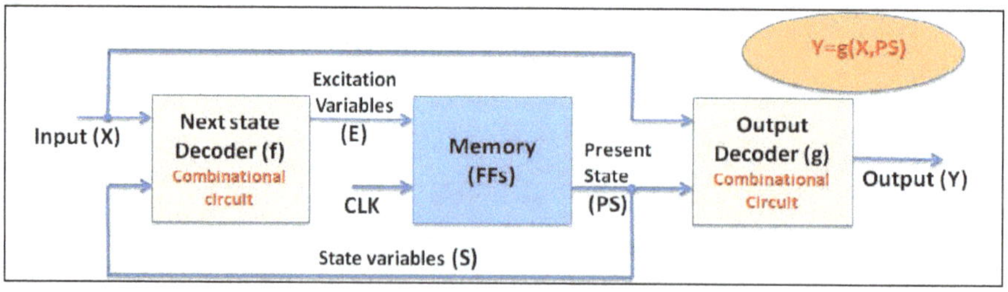

Class – A (Mealy) machine.

Class-A State Machine

The class A state machine is popularly known as Mealy machine, named after logic designer Mealy. It consists of three major blocks, next state decoder, memory and output decoder along with feedback and feed-forward mechanism. The basic property of the Mealy machine is that the output is a function of present input conditions (X) and the present state (PS).

Y= g(X,PS).

value of counter is decided by the external input. Another example of class D machine is the look-up memory.

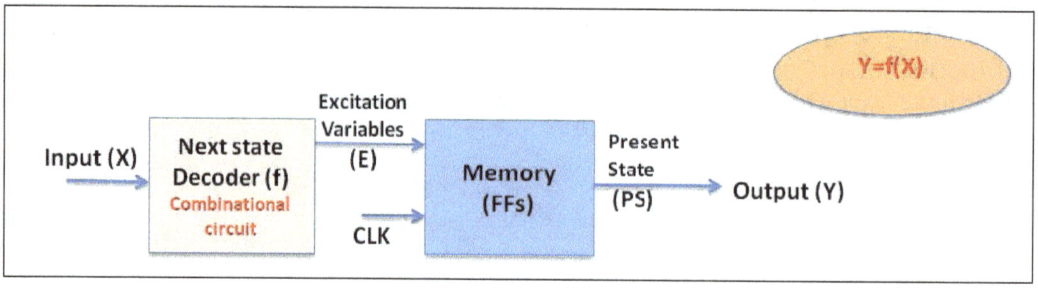

Class – D machine.

Class-E State Machine

The class E state machine is the simplest sequential circuit model, consisting of only the flip flops as the memory element without any combinational circuit. Ripple counter can be considered as the example of class-E state machine.

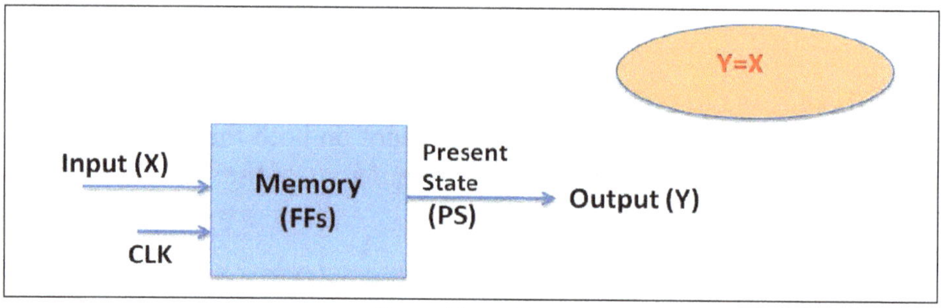

Class – E machine.

Behavior of Flip Flops

Role of Flip flops as memory is very important in the finite state machine. Before designing the state machine, it is important to look at the behavior of flip flop. The behavior f lip flops is can be uniquely described by its logic symbol, characteristic table, characteristics equation, excitation table, state diagram and state table.

SR Flip Flop

The characteristic table defines the state of each flip flop as a function of its S, R inputs and previous state. With reference to clock pulse, Q represents the present state and Q(t+1) refers to next state after the clock transition. The characteristics table for RS flip flop is prepared by writing Present state, S, R as inputs in the initial columns and the next state in the last column. First four rows assumes that Q=0 and Q=1 for next four rows. All possible input conditions for S and R are written for group of four rows.

The characteristics table for SR flip flop shows that next state is equal to the present state when both S and R inputs are equal to 0. When R=1, application of clock pulse resets the flip flop. When S=1, the flip flop sets to 1. When both S and R inputs are 1, the output state is unknown or un-deterministic. For such case, let us consider output as ? or X.

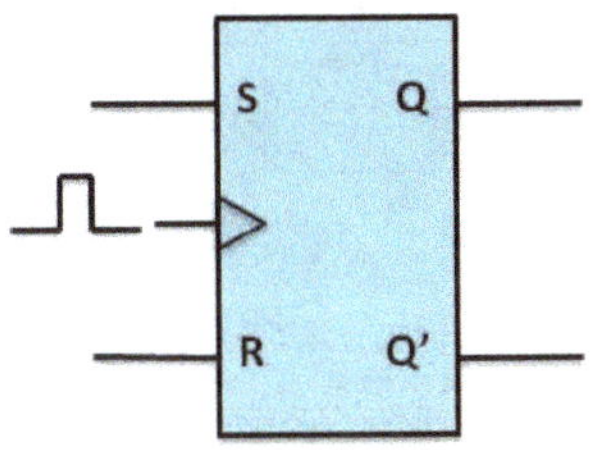

Characteristic Table

Q	S	R	Q(t+1)
0	0	0	0
0	0	1	0
0	1	0	1
0	1	1	X
1	0	0	1
1	0	1	0
1	1	0	1
1	1	1	X

K-map

Q\SR	00	01	11	10
0			X	1
1	1		X	1

Characteristic Equation

Q(t+1) = S+R'Q

Characteristics table and characteristics equations for SR flip flop.

Let us prepare a k-map of 3 variables Q, S and R and map the corresponding 1s and carry out simplification visually. Thus the characteristic equation for SR flip flop is,

$$Q(t+1) = S + R'Q$$

D Flip Flop

The characteristics table for D flip flop is prepared by writing present state, and D inputs in the initial columns and the next state in the last column. In the characteristic table, the next state of D flip flop is completely dependent on input D and independent of present state.

Let us map the characteristics table in 2-variable K-map and map the corresponding 1s and carry out simplification visually. Thus the characteristic equation for D flip flop is,

$$Q(t+1) = D$$

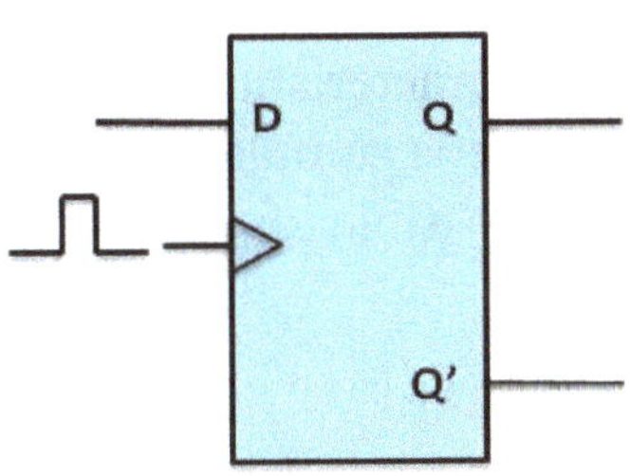

Characteristic Table

Q	D	Q(t+1)
0	0	0
0	1	1
1	0	0
1	1	1

K-map

Q\D	0	1
0		1
1		1

Characteristic Equation

Q(t+1) = D

Characteristics table and characteristics equations for D flip flop.

JK Flip Flop

The characteristics table for JK flip flop is prepared by writing Present state, JK inputs in the initial columns and the next state in the last column. The characteristics table for JK flip flop is similar to SR flip flop except the unknown condition. When both J and K are equal to 1, the next state is equal to the complement of present state i.e. Q(t+1)=Q'.

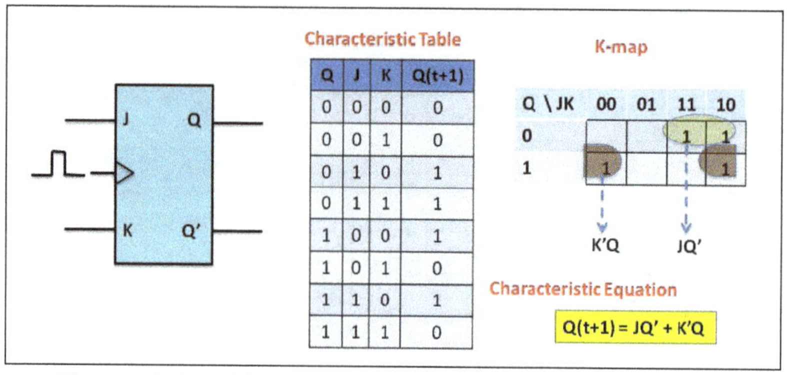

Q	J	K	Q(t+1)
0	0	0	0
0	0	1	0
0	1	0	1
0	1	1	1
1	0	0	1
1	0	1	0
1	1	0	1
1	1	1	0

Characteristics table and characteristics equations for JK flip flop.

Let us prepare a k-map of 3 variables Q, J and K and map the corresponding 1s and carry out simplification visually. Thus the characteristic equation for JK flip flop is,

$$Q(t+1) = JQ' + K'Q$$

T Flip Flop

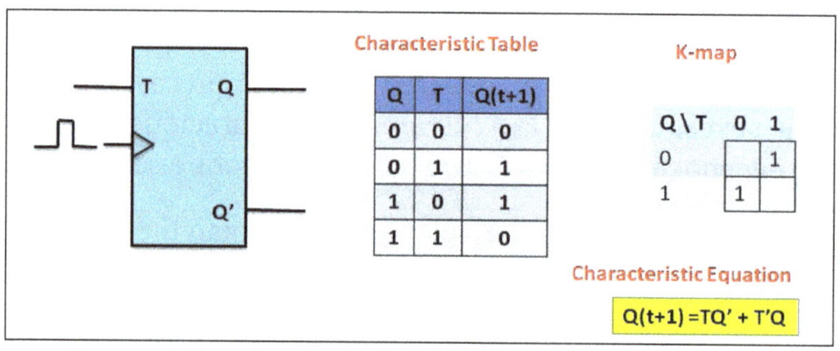

Q	T	Q(t+1)
0	0	0
0	1	1
1	0	1
1	1	0

Characteristics table and characteristics equations for T flip flop.

The characteristics table for T flip flop is prepared by writing present state and T inputs in the initial columns and the next state in the last column. In the characteristic table, the next state of T flip flop is same as the present state (Q) if T=0 and next state is complemented when T=1. Let us map the characteristics table in 2-variable K-map. It is not possible to simplify the K-map. The characteristic equation for T flip flop is given by,

$$Q(t+1) = TQ' + T'Q.$$

Excitation Table

During the design process, one of the important tasks is to find out the flip flop input conditions when the present state and next state transitions are known. For this reason excitation table is introduced to know the flip flop excitation inputs from the given states.

Table: Excitation table for SR flip flop.

Present State	Next State	Inputs	
Q	Q(t+1)	S	R

0	0	0	X
0	1	1	0
1	0	0	1
1	1	X	1

Let us now prepare the excitation table for SR flip flop. In order to obtain the excitation table of Flip flop, it is necessary to list all possible present state and next states and obtain the input conditions to generate such state change. If both present and next states are equal to 0, then SR inputs should be 0X. Here X mean don't care condition i.e. the input can be either 0 or 1. If present state is 0 and next state 1 is desired then SR inputs should be 10. Transition of present state 1 to next state 0 can be obtained by setting SR inputs to 01. Transition of present state 1 to next state 0 can be obtained by setting SR inputs to 01. The SR inputs should be X0 for 1 to 1 transition of present to next states.

Table: Excitation table for JK flip flop.

Present State	Next State	Inputs	
Q	Q(t+1)	J	K
0	0	0	X
0	1	1	X
1	0	X	1
1	1	X	0

To obtain the excitation table of JK Flip flop, it is necessary to list all possible present state and next states to determine the input conditions to generate such state change. If both present and next states are equal to 0, then JK inputs should be 0X. Here X stand for don't care condition i.e. the input can be either 0 or 1. If present state is 0 and next state 1 is desired then JK inputs should be 1X. Transition of present state 1 to next state 0 can be obtained by setting JK inputs to X1. The JK inputs should be X0 for 1 to 1 transition of present to next states. The JK flip flop is normally preferred for almost all designs because the excitation table indicates do not care (X) input either at J or K input for all present state to next state. This helps in eliminating many terms from the next state equations.

Table: Excitation table for D flip flop.

Present State	Next State	Inputs
Q	Q(t+1)	D
0	0	0
0	1	1
1	0	0
1	1	1

To obtain the excitation table of D Flip flop, it is necessary to list all possible present state and next states to determine the input conditions to generate such state change. If both present and next states are equal to 0, then D input should be 0. If present state is 0 and next state 1 is desired then

D input should be 1. Transition of present state 1 to next state 0 can be obtained by setting D input to 0. The D input should be 1 for 1 to 1 transition of present to next states. To obtain the excitation table of T Flip flop, it is necessary to list all possible present state and next states to determine the input conditions to generate such state change.

Table: Excitation table for T flip flop.

Present State	Next State	Inputs
Q	Q(t+1)	T
0	0	0
0	1	1
1	0	1
1	1	0

If both present and next states are equal to 0, then T input should be 0. If present state is 0 and next state 1 is desired then T input should be 1. Transition of present state 1 to next state 0 can be obtained by setting T input to 1. The D input should be 0 for 1 to 1 transition of present to next states.

State Diagrams

In the sequential circuits, he output of the circuit at any instant of time depends upon its present state and the inputs. These also determine the next state of the circuit. The relationship between inputs, outputs, present state and next state can be specified by either the state table or the state diagram.

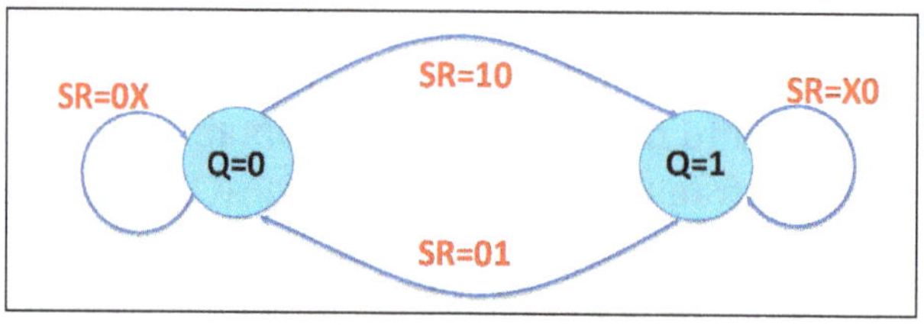

State diagram of SR flip flop.

A state is represented by a circle. There are two states for all flip flops. Flip flop can be in Reset state when Q=0 and in the Set state when Q=1. From the excitation table, it is clear that all the flip flops can have 4 possible transitions. The transition between the states is indicated by directed lines (or arcs) connecting the circle. Two of them are referred to as self-transitions and rest two is state transitions.

If SR flip flop is in Reset state (Q=0) and SR inputs are 0X, flip flop remains in the same state. Here X value can be either 0 or 1 i.e. don't care condition. But if SR inputs are 10 then FF makes 0 to 1 state transition. Similarly when flip flop is in Set state (Q=1) and SR inputs are X0, and then flip flop remains in the same state. But if SR inputs are 01, then FF makes 1 to 0 state transitions.

Let us now prepare state diagram for JK flip flop. Even JK flip flop has two states - Reset and Set state. Let us prepare state diagram with the help of excitation table. If JK flip flop is in Reset state (Q=0) and JK inputs are 0X, flip flop remains in the same state. But if JK inputs are 1X then FF makes 0 to 1 state transition. Similarly when flip flop is in Set state (Q=1) and JK inputs are X0, then flip flop remains in the same state. But if JK inputs are X1, then FF makes 1 to 0 state transitions.

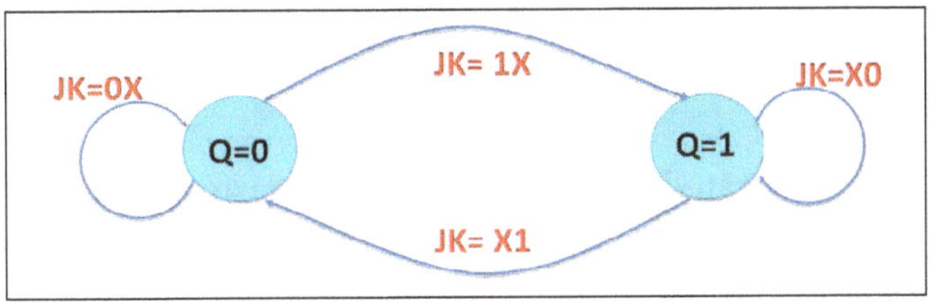

State diagram of JK flip flop.

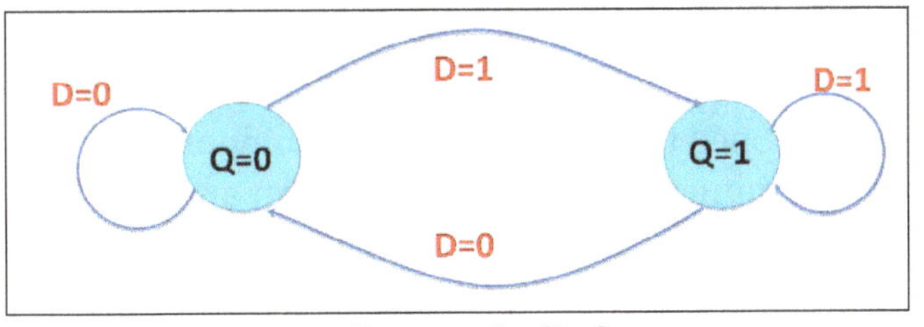

State diagram of D flip flop.

Let us now prepare state diagram for D flip flop. There are two states - Reset state when Q=0 and in the Set state when Q=1. If D flip flop is in Reset state (Q=0) and D input is 0, flip flop remains in the same state. But if D input is 1 then FF makes 0 to 1 state transition. When flip flop is in Set state (Q=1) and D input is 1, then flip flop remains in the same state. But if D input is 0, then FF makes 1 to 0 state transitions.

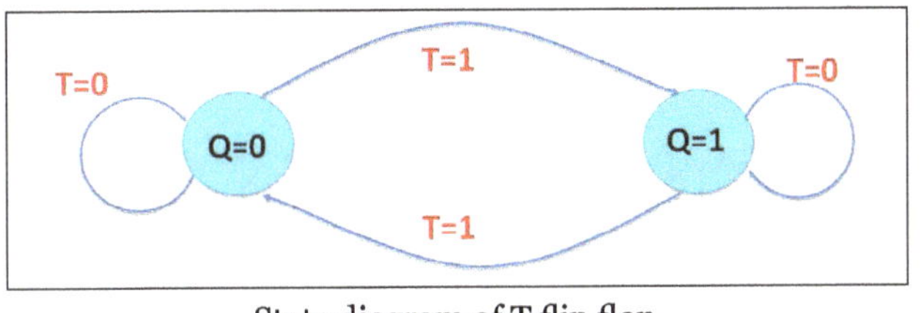

State diagram of T flip flop.

Let us now prepare state diagram for T flip flop. There are two states - Reset state when Q=0 and in the Set state when Q=1. If T flip flop is in Reset state (Q=0) and T input is 0, flip flop remains in the same state. But if T input is 1 then FF makes 0 to 1 state transition. When flip flop is in Set state (Q=1) and T input is 0, then flip flop remains in the same state. But if T input is 1, then FF makes 1 to 0 state transitions.

References

- Digital-electronics: electronics.fandom.com, Retrieved 14, August 2020
- Digital-signal: chegg.com, Retrieved 21, January 2020
- Analogdigitalsignals: lancasterschools.org, Retrieved 02, June 2020
- Advantages-and-disadvantages-of-digital-signals: geeksforgeeks.org, Retrieved 13, February 2020
- Digital-integrated-circuits: wellpcb.com, Retrieved 28, July 2020

2

Logic Families

The collection of different integrated circuit chips which perform different logic gate functions like AND, OR, NOT but have similar input, output and integrated circuit characteristics are termed as logic families. A few of the major types of logic families are resistor-transistor logic, diode-transistor logic, emitter-coupled logic, transistor-transistor logic and MOS logic family. This chapter will explain briefly all these divisions of logic families.

The first logic circuit was developed using discrete circuit components. Using advance techniques, these complex circuits can be miniaturized and produced on a small piece of semiconductor material like silicon. Such a circuit is called integrated circuit (IC). Now-a-days, all the digital circuits are available in IC form. While producing digital ICs, different circuit configurations and manufacturing technologies are used. This results into a specific logic family. Each logic family designed in this way has identical electrical characteristics such as supply voltage range, speed of operation, power dissipation, noise margin etc.

Significance of Logic Families

Almost all electronic gadgets make use of different digital systems for their operation. All the digital systems use some kind of digital ICs. For the sake of simplicity in design and compatibility in constructing any complex digital system, all digital circuits (ICs) used in the design process should be from same logic family.

For the expansion of the system, it is necessary to connect different logic circuits together. In order to connect the output of one logic circuit to the input of another logic circuit, one must have circuit with similar characteristics. If the electrical IO characteristics of these logic circuits are not similar, there is a need to design an interfacing circuit to maintain the compatibility of digital logic ICs. This interfacing circuit will match the electrical characteristics of the logic circuits. This ensures the compatibility for proper operation of the circuit. This is the key concept behind family for logic circuits. In a logic family, the family members have similar electrical characteristics. Digital logic circuit has to be designed considering these compatibilities of different logic families in terms of different characteristics and parameters associated with the families.

Types of Logic Families

Logic families are the logic circuits having identical electrical parameters. It is a group of compatible ICs with the same logic levels and supply voltages for performing various logic functions. They are fabricated using a specific circuit configuration which is referred to as a Logic family. The circuit design of the basic gate of each logic family is the same. The logic family is designed by considering the basic electronic components such as resistors, diodes, transistors, and MOSFET; or combinations of any of these components. Accordingly, logic families are classified as per the construction of the basic logic circuits. Many different logic families of digital ICs have been introduced commercially are listed in table.

Table: Logic families and the components used for construction of logic family.

Name of logic family	Components used
DL(Diode Logic)	Diodes
RTL(Resistor Transistor Logic)	Resistors and transistors
DTL(Diode Transistor Logic)	Diodes, transistors and resistors
TTL(Transistor Transistor Logic)	Transistors and resistors
ECL(Emitter Coupled Logic)	Transistors and diodes
PMOS(P channel Metal Oxide Semiconductor Logic)	P-MOSFETs
NMOS(N channel Metal Oxide Semiconductor Logic)	N-MOSFETs
CMOS(Complementary Metal Oxide Semiconductor Logic)	P–MOSFET and N-MOSFET

Logic families are classified according to the principle type of electronic components used in their circuitry as shown in Figure. They are:

- Bipolar ICs which uses diodes and transistors (BJT).
- Unipolar ICs which uses MOSFETs.

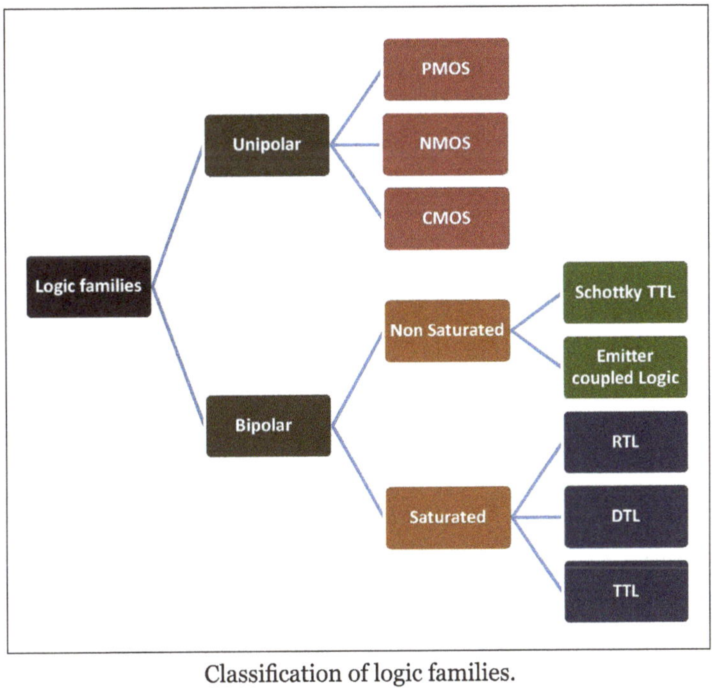

Classification of logic families.

Out of the logic families mentioned above, DL, RTL and DTL are not very useful due to some inherent disadvantages while TTL, ECL and CMOS are widely used in many digital circuit design applications. Each logic family is further classified based on the significant variation in the key parameters. This helps in improving the performance of that logic family. For example, TTL logic family has subfamilies such as low power TTL, Schottky TTL etc.

TTL has an extensive list of digital functions and is currently the most popular logic family. ECL is used in systems requiring high-speed operations. MOS is used in circuits requiring a high component density, whereas CMOS is used in systems requiring low power consumption. These three families, TTL, ECL and CMOS have LSI devices and also a large number of MSI and SSI devices. SSI devices are those that come with a small number of gates or flip flops in one IC package. The limit on the number of circuits in SSI devices is number of pins in a package. A 14 pin package can accommodate only four two input gates because, each gate requires three external pins-two each for inputs and one each for output, for a total of 12 pins. The remaining two pins are needed for supplying power to the circuits.

TTL ICs are usually distinguished by numerical designations such as the 5400 and 7400 series. The former has wide operating temperature range and is suitable for military use and the latter has a narrower temperature range and is suitable for industrial use only. TTL SSI ICs such as 7404, 7400 etc. are available in the market.

The most common ECL types are designated as the 10000 series. The 10102 provides four two input NOR gates. Note that the ECL gate may have two outputs, one for the NOR function and the other for the OR function. ECL gates have three terminals for power supply. VCC1 and VCC2 are usually connected to ground and VEE to a -5.2 Volt supply.

CMOS circuits are designated by the 4000 series. Only two NOR gates can be accommodated in the 4002 because of limited number of pins. The 4050 type provides six buffer gates. Both ICs have two unused terminals marked NC. The terminal marked Vdd requires a power supply voltage from 3 to 15 Volt while Vss is usually connected to ground.

Positive and Negative Logic

A logic level is one of the several states a digital circuit can have. It is expressed in DC voltage with reference to ground. Usually, this term refers to binary logic level consisting of two states viz. low state or 0V or logic 0 and high state or +V or logic high. In addition to two state digital circuits, some circuits are three state or tri-state logic circuits. Third state or high impedance state is included in addition to two logic states i.e. logic 0 and logic 1. If the circuit is in this state, it is effectively cancelling the output. Every logic gate accepts inputs and produces output in terms of logic and voltage levels. It is possible to interpret logic and voltage levels in terms of two types of logic configurations: positive logic and negative logic. Table shows the two assignments that define positive and negative logic systems.

Table: Positive and Negative logic assignments.

Positive logic	Negative logic
H = 1	H = 0
L = 0	L = 1

The binary signal at the input or output of the any gate has one of two values except during transition. One signal value represents logic 1 and other as logic 0. Since two signal values are assigned to two logic values, there exist two different assignments of signal to logic. Due to the principle of duality of Boolean algebra, an interchange of signal values assignment results in a dual function implementation.

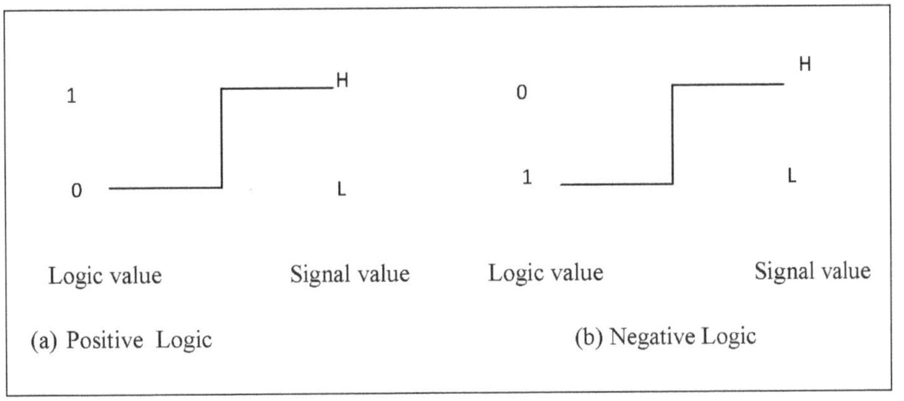

Logic level assignments.

Consider the two values of a binary signal as shown in Figure. One value must be higher than the other since the two values must be different in order to be distinct the higher level is designated as H and the lower level as L. There are two choices of logic value assignment:

- Choosing the high level H to represent logic 1defines positive logic.
- Choosing the low level L to represent logic 1 defines negative logic.

It is not the signal polarity that determines the type of logic; but rather the assignment of logic values according to the relative amplitudes of the signal. Integrated circuit data sheets define digital function in terms of H and L and not in terms of logic 0 or logic 1. It is up to the user to determine the positive and negative assignments. In each family, there is a range of voltage values that the circuit will recognize as a high or low level. Consider table below, to understand the logical interpretation of any gate. It is depends on the convention of positive or negative logic.

Table: Logical interpretation of logic gate behavior.

Logical implementation		Physical implementation		Physical implementation	
Gate Behaviour		Positive logic		Negative logic	
A B	Y	A B	AND	A B	OR
L L	L	0 0	0	1 1	1
L H	L	0 1	0	1 0	1
H L	L	1 0	0	0 1	1
H H	L	1 1	1	0 0	0

This table demonstrates how physical implementation of logic gate, changes with positive and negative logic implementations. A physical implementation of gate has two different representations depending on whether positive or negative logic is used. Positive logic inputs and outputs are also known as Active High signals. Negative logic inputs and outputs are also known as Active Low signals. Input and output polarity may be indicated by bubble or no bubble. Figure indicates two different implementations namely positive logic AND gate and a negative logic OR gate for the same logic level implementation.

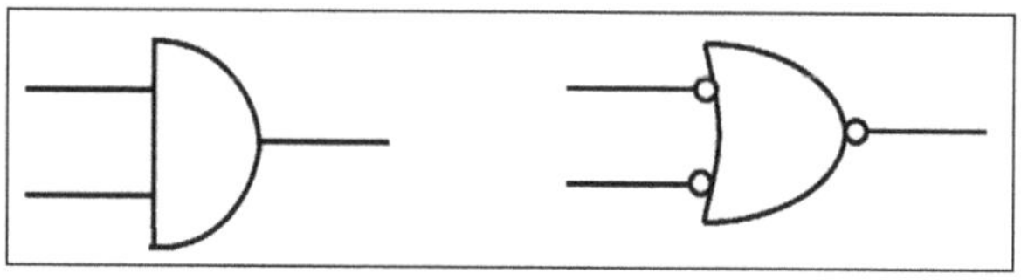

Characteristics of Logic Families

Here, we discuss the different parameters which are used to characterize different logic families:

1. HIGH-level input current, I_{IH}: This is the current flowing into (taken as positive) or out of (taken as negative) an input, when a HIGH-level input voltage equal to the minimum HIGH-level output voltage specified for the family is applied. In the case of bipolar logic families such as TTL, the circuit design is such that this current flows into the input pin and is therefore specified as positive. In the case of CMOS logic families, it could be either positive or negative, and only an absolute value is specified in this case.

2. LOW-level input current, I_{IL}: The LOW-level input current is the maximum current flowing into (taken as positive) or out of (taken as negative) the input of a logic function, when the voltage applied at the input equals the maximum LOW-level output voltage specified for the family. In the case of bipolar logic families such as TTL, the circuit design is such that this current flows out of the input pin and is therefore specified as negative. In the case of CMOS logic families, it could be either positive or negative. In this case, only an absolute value is specified.

3. HIGH-level output current, I_{OH}: This is the maximum current flowing out of an output when the input conditions are such that the output is in the logic HIGH state. It is normally shown as a negative number. It informs us about the current sourcing capability of the output. The magnitude of I_{OH} determines the number of inputs the logic function can drive when its output is in the logic HIGH state. For example, for the standard TTL family, the minimum guaranteed I_{OH} is −400 μA, which can drive 10 standard TTL inputs with each requiring 40 μA in the HIGH state.

4. LOW-level output current, I_{OL}: This is the maximum current flowing into the output pin of a logic function when the input conditions are such that the output is in the logic LOW state. It informs us about the current sinking capability of the output. The magnitude of IOL determines the number of inputs the logic function can drive when its output is in the logic LOW state. For example, for the standard TTL family, the minimum guaranteed IOL is 16 mA, which can drive 10 standard TTL inputs with each requiring 1.6mA in the LOW state.

5. HIGH-level input voltage, V_{IH}: This is the minimum voltage level that needs to be applied at the input to be recognized as a legal HIGH level for the specified family. For the standard TTL family, a 2 V input voltage is a legal HIGH logic state.

6. LOW-level input voltage, V_{IL}: This is the maximum voltage level applied at the input that is recognized as a legal LOW level for the specified family. For the standard TTL family, an input voltage of 0.8 V is a legal LOW logic state.

7. HIGH-level output voltage, V_{OH}: This is the minimum voltage on the output pin of a logic function when the input conditions establish logic HIGH at the output for the specified family. In the case of the standard TTL family of devices, the HIGH level output voltage can be as low as 2.4V and

still be treated as a legal HIGH logic state. It may be mentioned here that, for a given logic family, the VOH specification is always greater than the VIH specification to ensure output-to-input compatibility when the output of one device feeds the input of another.

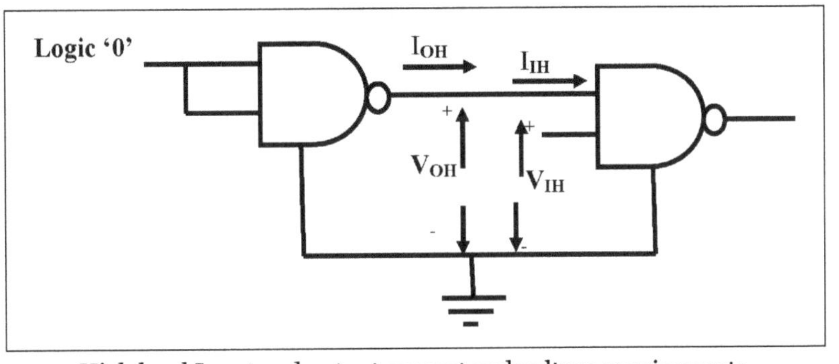

High level Input and output current and voltage requirements.

8. LOW-level output voltage, V_{OL}: This is the maximum voltage on the output pin of a logic function when the input conditions establish logic LOW at the output for the specified family. In the case of the standard TTL family of devices, the LOW-level output voltage can be as high as 0.4V and still be treated as a legal LOW logic state. It may be mentioned here that, for a given logic family, the VOL specification is always smaller than the VIL specification to ensure output-to-input compatibility when the output of one device feeds the input of another.

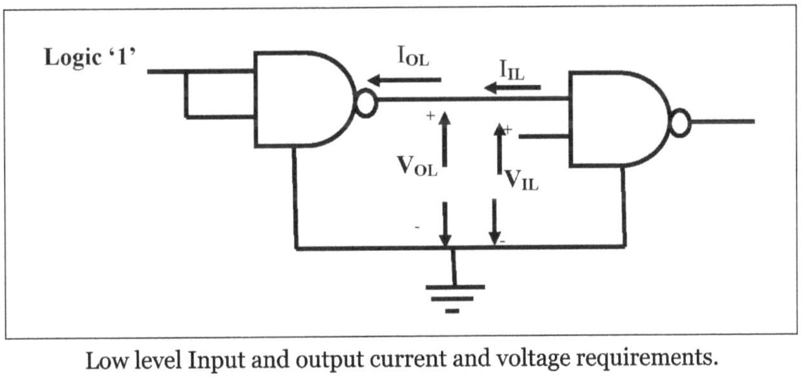

Low level Input and output current and voltage requirements.

9. Supply current, I_{CC}: The supply current when the output is HIGH, LOW and in the high-impedance state is respectively designated as I_{CCH}, I_{CCL} and I_{CCZ}.

10. Fan out: It specifies the number of standard loads that the output of the gate can drive without affecting its normal operation. A standard load is usually defined as the amount of current needed by an input of another gate in the same family as shown in Figure. Sometimes, the term' loading' is also used instead of 'fan out'. This term is derived from the fact that the output of the gate can supply a limited amount of current above which it ceases to operate properly and is said to be overloaded. The output of the gate is usually connected to the inputs of similar gates.

Each input consumes a certain amount of power from gate input so that, each additional connection adds to load the gate. These loading rules are listed for family of standard digital circuits. The rules specify the maximum amount of loading allowed for each output in the circuit. Exceeding the specified maximum load may cause a malfunction because the circuit cannot supply the power demanded from it. Thus fan out is the maximum number of inputs that can be connected to the output of the gate and is expressed by a number. The fanout capabilities of a gate must be considered while simplifying Boolean functions. One should use non inverting amplifiers or buffers to provide additional driving capabilities for heavy loads.

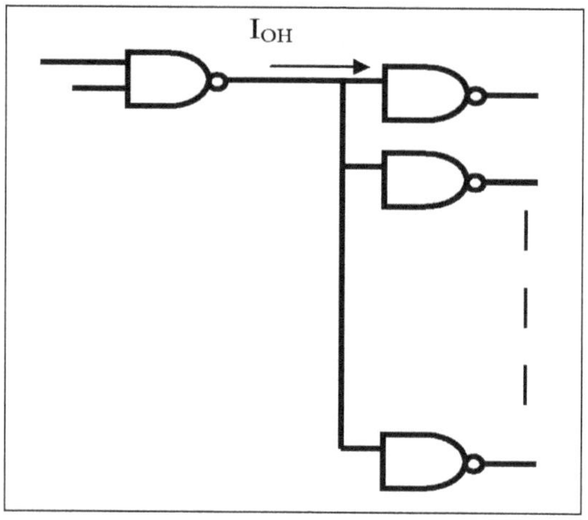

Fan out of logic gate is decided by number of loads connected to output of gate of same logic family.

11. Fan in: This is the number of inputs of a logic gate. It is decided by the input current sinking capability of a logic gate as shown in Figure.

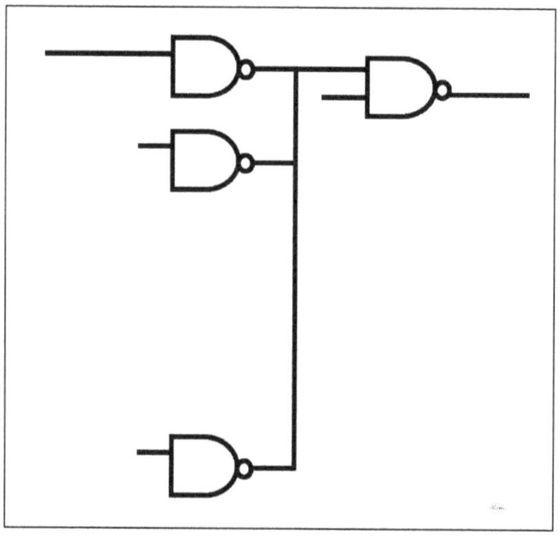

Fan in of logic gate.

12. Power Dissipation: This is the power supplied required to operate the gate. It is expressed in milli-watt (mW) and represents actual power dissipated in the gate. It is the number that represents power delivered to gate from the power supply. The total power dissipated in the digital system is sum of power dissipated in each digital IC.

13. Rise time, t_r: This is the time that elapses between 10% and 90 % of the final signal level when the signal makes a transition from logic LOW to logic HIGH.

14. Fall time, t_f: This is the time that elapses between 90 and 10 % of the signal level when the signal makes a transition from logic LOW to logic HIGH.

15. Propagation delay t_p: The propagation delay is the time delay between the occurrences of change in the logical level at the input and before it is reflected at the output. It is the time delay between the specified voltage points on the input and output waveforms. Propagation delays are separately defined for LOW-to-HIGH and HIGH-to-LOW transitions at the output. In addition, we also define enable and disable time delays that occur during transition between the high-impedance state and defined logic LOW or HIGH states.

- Propagation delay t_{pLH}: This is the time delay between the specified voltage points on the input and output waveforms with the output changing from LOW to HIGH.
- Propagation delay t_{pHL}: This is the time delay between the specified voltage points on the input and output waveforms with the output changing from HIGH to LOW.

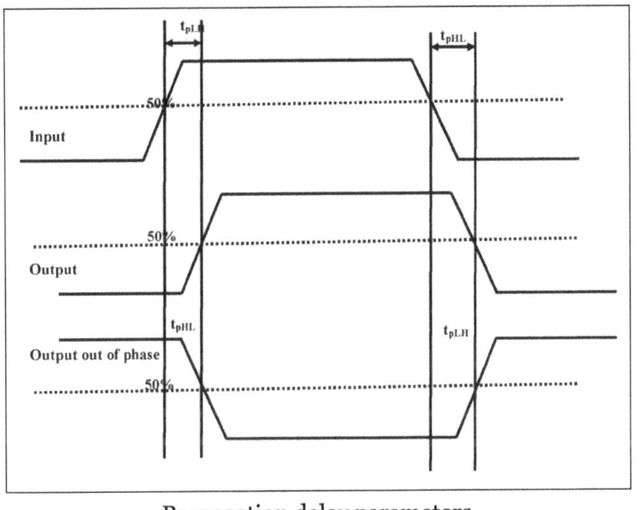

Propagation delay parameters.

It is the average transition delay time for a signal to propagate from input to output when the binary signals change in value. The signal through gate takes a certain amount of time to propagate from the inputs to the output. The interval of time is defined as the propagation delay of the gate. Propagation delay is expressed in nanoseconds(ns)(1ns = 10^{-9} s). The signals travelling from input to output of the system pass through a number of gates. The propagation delay of the system is the sum of the propagation delays of all these gates. When the speed of operation is important, each gate must have a small propagation delay and the digital circuit must have minimum number of gates between input and output.

16. Disable time from the HIGH state, t_{pHZ}: Defined for a tristate device, this is the time delay between the specified voltage points on the input and output waveforms with the tristate output changing from the logic HIGH level to the high-impedance state.

17. Disable time from the LOW state, t_{pLZ}: Defined for a tristate device, this is the time delay between the specified voltage points on the input and output waveforms with the tristate output changing from the logic LOW level to the high-impedance state.

18. Enable time from the HIGH state, t_{pZH}: Defined for a tristate device, this is the time delay between the specified voltage points on the input and output waveforms with the tristate output changing from the high-impedance state to the logic HIGH level.

19. Power dissipation: The power dissipation parameter for a logic family is specified in terms of power consumption per gate and is the product of supply voltage V_{CC} and supply current I_{CC}. The supply current is taken as the average of the HIGH-level supply current I_{CCH} and the LOW-level supply current I_{CCL}.

20. Speed–power product: The speed of a logic circuit can be increased, that is, the propagation delay can be reduced, at the expense of power dissipation. We may recall that, when a bipolar transistor switches between cut-off and saturation, it dissipates the least power, but has a large associated switching time delay. On the other hand, when the transistor is operated in the active region, power dissipation goes up, while the switching time decreases drastically. It is always desirable to have low values for both propagation delay, and power dissipation parameters. A useful figure-of-merit used to evaluate different logic families is the speed–power product, expressed in picojoules, which is the product of the propagation delay (measured in nanoseconds) and the power dissipation per gate (measured in milliwatts).

21. Noise margin: This is the maximum noise voltage added to the input signal of digital circuit that does not cause an undesirable change in the circuit output. There are two types of noise to be considered here:

- DC noise: This is caused by a drift in the voltage levels of a signal.
- AC noise: This is caused by random pulse that may be created by other switching signals.

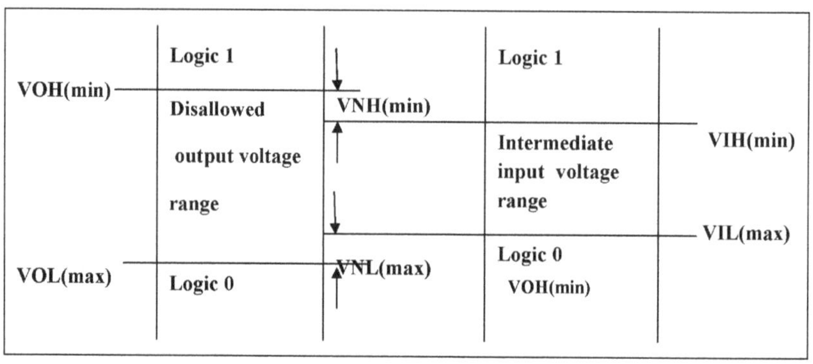

This is a quantitative measure of noise immunity offered by the logic family. When the output of a logic device feeds the input of another device of the same family, a legal HIGH logic state at the output of the feeding device should be treated as a legal HIGH logic state by the input of the device being fed. Similarly, a legal LOW logic state of the feeding device should be treated as a legal LOW logic state by the device being fed. The legal HIGH and LOW voltage levels for a given logic family are different for outputs and inputs. Figure shows the generalized case of legal HIGH and LOW voltage levels for output. As we can see from the two diagrams, there is a disallowed range of output voltage levels from VOL(max.) to VOH(min.) and an indeterminate range of input voltage levels from VIL(max.) to VIH(min.). Since VIL(max.) is greater than VOL(max.), the LOW output state can tolerate a positive voltage spike equal to (VIL(max.) – VOL(max)); and still be a legal LOW input. Similarly, VOH(min.) is greater than VIH (min.), and the HIGH output state can tolerate a negative voltage spike equal to (VOH(min.) – VIH (min.)) and still be a legal HIGH input. Here, (VIL(max.) – VOL(max.)) and (VOH(min.) – VIH (min.)) are respectively known as the LOW-level and HIGH-level noise margin.

Resistor-Transistor Logic

Resistor-Transistor Logic, or RTL, refers to the obsolete technology for designing and fabricating digital circuits that employ logic gates consisting of nothing but transistors and resistors. RTL gates are now seldom used, if at all, in modern digital electronics design because it has several drawbacks, such as bulkiness, low speed, limited fan-out, and poor noise margin. A basic understanding of what RTL is, however, would be helpful to any engineer who wishes to get familiarized with TTL, which for the past many years has become widely used in digital devices such as logic gates, latches, buffers, counters, and the like.

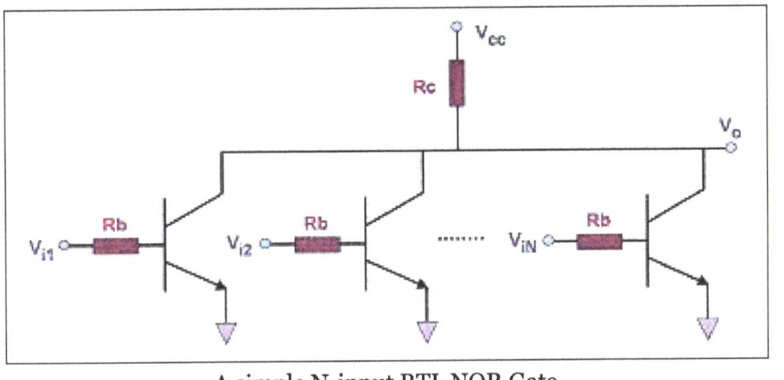

A simple N-input RTL NOR Gate.

Figure shows an example of an N-input RTL NOR gate. It consists of N transistors, whose collectors are all tied up to V_{cc} through a common resistor, and whose emitters are all grounded. Their bases individually act as inputs for input voltages V_i (i = 1,2,...,N), which represent input logic levels. The output V_o is taken across the collector- resistor node and ground. V_o is only 'high' if the inputs to the bases of all the transistors are 'low'.

One of the earliest gates used in integrated circuits is a special type of RTL gate known as the direct-coupled transistor logic (DCTL) gate. A DCTL gate is one wherein the bases of the transistors are connected directly to inputs without any base resistors. Thus, the RTL NOR gate shown in Figure becomes a DCTL NOR gate if all the base resistors (Rb's) are eliminated. Without the base resistors, DCTL gates are more economical and simpler to fabricate onto integrated circuits than RTL gates with base resistors.

The main drawback of DCTL gates is that they suffer from a phenomenon known as current hogging. Ideally, several transistors that are connected in parallel will share the load current equally among themselves when they are all brought into saturation. In the real world, however, the saturation points of different transistors are attained with different levels of input voltages to the base (V_{be}). As such, transistors that are in parallel and share the same input voltage (which are commonly encountered in DCTL circuits) do not share the load current evenly among themselves.

In fact, once the transistor with the lowest Vbesat saturates, the other transistors are prevented from saturating themselves. This causes the saturated transistor to 'hog' the load current, i.e., it carries the bulk of the load current whereas those transistors that were prevented from saturating carries a minimal portion of it. Current hogging, which prevented DCTL from becoming widely used, is largely avoided in RTL circuits simply by retaining the base resistors.

RTL gates also exhibit limited 'fan-outs'. The fan-out of a gate is the ability of its output to drive several other gates. The more gates it can drive, the higher is its fan-out. The fan-out of a gate is limited by the current that its output can supply to the gate inputs connected to it when the output is at logic '1', since at this state it must be able to drive the connected input transistors into saturation.

Another weakness of an RTL gate is its poor noise margin. The noise margin of a logic gate for logic level '0', Δ0, is defined as the difference between the maximum input voltage that it will recognize as a '0' (V_{il}) and the maximum voltage that may be applied to it as a '0' (V_{ol} of the driving gate connected to it). For logic level '1', the noise margin Δ1 is the difference between the minimum input voltage that may be applied to it as a '1' (V_{oh} of the driving gate connected to it) and the minimum input voltage that it will recognize as a '1' (V_{ih}). Mathematically, $\Delta 0 = V_{il}-V_{ol}$ and $\Delta 1 = V_{oh}-V_{ih}$. Any noise that causes a noise margin to be overcome will result in a '0' being erroneously read as a '1' or vice versa. In other words, noise margin is a measure of the immunity of a gate from reading an input logic level incorrectly.

In an RTL circuit, the collector output of the driving transistor is directly connected to the base resistor of the driven transistor. Circuit analysis would easily show that in such an arrangement, the differences between V_{il} and V_{ol}, and between V_{oh} and V_{ih}, are not that large. This is why RTL gates are known to have poor noise margins in comparison to DTL and TTL gates.

Diode Transistor Logic

Diode Transistor Logic, or DTL, refers to the technology for designing and fabricating digital circuits wherein logic gates employ both diodes and transistors. DTL offers better noise margins and greater fan-outs than RTL, but suffers from low speed, especially in comparison to TTL.

RTL allows the construction of NOR gates easily, but NAND gates are relatively more difficult to get from RTL. DTL, however, allows the construction of simple NAND gates from a single transistor, with the help of several diodes and resistors. Figure shows an example of an 3-input DTL NAND gate. It consists of a single transistor Q configured as an inverter, which is driven by a current that depends on the inputs to the three input diodes D_1-D_3.

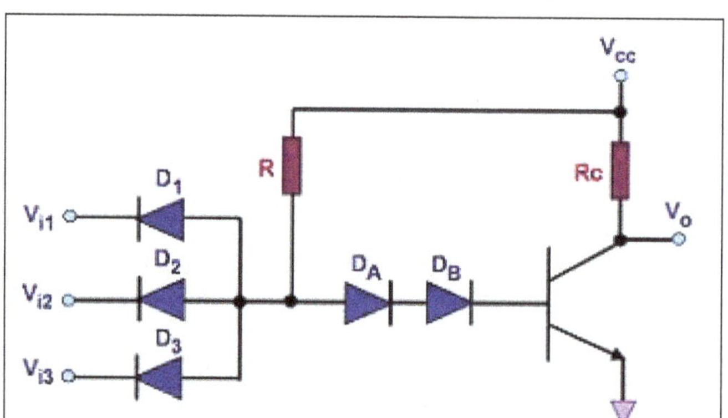

A simple 3-input DTL NAND Gate.

In the NAND gate in Figure, the current through diodes D_A and D_B will only be large enough to drive the transistor into saturation and bring the output voltage Vo to logic '0' if all the input diodes D_1-D_3 are 'off', which is true when the inputs to all of them are logic '1'. This is because when D_1-D_3 are not conducting, all the current from V_{cc} through R will go through D_A and D_B and into the base of the transistor, turning it on and pulling V_o to near ground.

However, if any of the diodes D_1-D_3 gets an input voltage of logic '0', it gets forward-biased and starts conducting. This conducting diode 'shunts' almost all the current away from the reverse-biased D_A and D_B, limiting the transistor base current. This forces the transistor to turn off, bringing up the output voltage Vo to logic '1'.

One advantage of DTL over RTL is its better noise margin. The noise margin of a logic gate for logic level '0', Δ0, is defined as the difference between the maximum input voltage that it will recognize as a '0' (V_{il}) and the maximum voltage that may be applied to it as a '0' (Vol of the driving gate connected to it). For logic level '1', the noise margin Δ1 is the difference between the minimum input voltage that may be applied to it as a '1' (V_{oh} of the driving gate connected to it) and the minimum input voltage that it will recognize as a '1' (V_{ih}).

Mathematically, $\Delta 0 = V_{il} - V_{ol}$ and $\Delta 1 = V_{oh} - V_{ih}$. Any noise that causes a noise margin to be overcome will result in a '0' being erroneously read as a '1' or vice versa. In other words, noise margin is a measure of the immunity of a gate from reading an input logic level incorrectly.

In a DTL circuit, the collector output of the driving transistor is separated from the base resistor of the driven transistor by several diodes. Circuit analysis would easily show that in such an arrangement, the differences between V_{il} and V_{ol}, and between V_{oh} and V_{ih}, are much larger than those exhibited by RTL gates, wherein the collector of the driving transistor is directly connected to the base resistor of the driven transistor. This is why DTL gates are known to have better noise margins than RTL gates.

One problem that DTL doesn't solve is its low speed, especially when the transistor is being turned off. Turning off a saturated transistor in a DTL gate requires it to first pass through the active region before going into cut-off. Cut-off, however, will not be reached until the stored charge in its base has been removed. The dissipation of the base charge takes time if there is no available path from the base to ground. This is why some DTL circuits have a base resistor that's tied to ground, but even this requires some trade-offs. Another problem with turning off the DTL output transistor is the fact that the effective capacitance of the output needs to charge up through Rc before the output voltage rises to thc final logic '1' level, which also consumes a relatively large amount of time. TTL, however, solves the speed problem of DTL elegantly.

Emitter-Coupled Logic

Emitter-coupled logic is the fastest of all digital logic families. It was invented by Hannon S. Yourke in the year 1956 at IBM. It is also called as current mode logic. The design of ECL circuit consists of transistors and resistors. By preventing the transistor from entering into saturation, the high-speed operation is achieved in ECL logic family. Very small voltage swing is necessary to switch between the two different voltage levels. This cannot be achieved in transistor-transistor logic, as the transistors enter into saturation mode, while in operation. Emitter-coupled logic family offers an incredible propagation delay of 1ns. The delay is more reduced in the latest ECL families.

Inverter Circuit of Emitter-coupled Logic

The circuit shown below represents the emitter-coupled logic circuit of an inverter. It has two NPN transistors connected in differential single-ended input mode. Both the emitters are connected together with common resistance R_E. It is a current limiting resistance, used to prevent the transistor from entering into saturation.

It has two outputs: inverting output(V_{OUT1}) and non-inverting output(V_{OUT12}). V_{IN} is the input terminal, where LOW or HIGH input is given. When the input is HIGH, it will turn ON the transistor Q_1 but not saturated and the transistor Q_2 is turned OFF. This will pull the output V_{OUT2} to HIGH but due to the drop in resistant R_1, the output at terminal V_{OUT1} will be at LOW value.

On the other side, when the input V_{IN} is given LOW value, it will turn OFF the transistor Q_1 and the transistor is turned ON. The transistor Q_2 will not enter into saturation. It will make the output at terminal V_{OUT1} to be pulled HIGH value. Due to the drop in resistance R_2, the output at terminal V_{OUT2} will have LOW value.

Two Input ECL OR/NOR Gate

The following circuit is the Emitter-coupled logic circuit of the 2-input OR/NOR gate. It is the slight modification of the inverter circuit given above. In this, an additional transistor is used at the input side. The operation is simple as explained above. If the input at both the transistors Q_1 and Q_2 are LOW, it will make V_{OUT1} to HIGH value. It corresponds to the NOR gate output. At the same time, transistor Q_3 is turned ON, which will make the V_{OUT2} to be HIGH. It corresponds to the OR gate output.

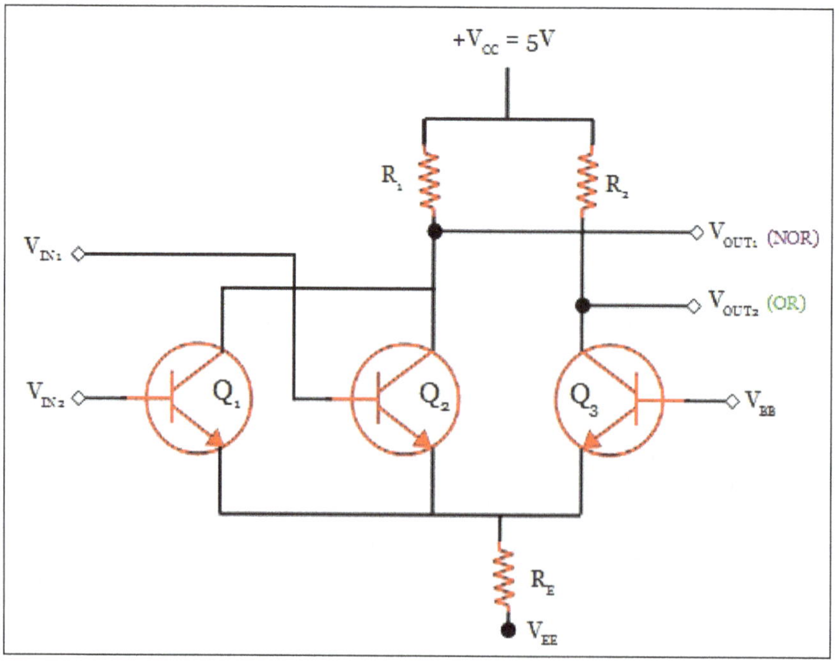

Similarly, if both the input of transistors Q_1 and Q_2 are HIGH, it will turn on both the transistors.

It will drive the output at terminal V_{OUT1} to be LOW. The transistor Q_3 is turned OFF during this operation. It will drive the output at terminal V_{OUT2} to be HIGH. The truth table for OR/NOR gate is shown below:

Inputs		OR	NOR
A	B	Y	Y
0	0	0	1
0	1	1	0
1	0	1	0
1	1	1	0

Advantages

- High-speed operation is possible and so the fastest logic family.
- Since transistors are not allowed to enter into saturation, which reduces the storage delay.
- Fan-out capability is high.

Apart from the advantages, it also has its own disadvantage. For the fast switching of transistors, the low and high logic levels are kept close. It reduces the noise margin. Since transistors are not allowed to enter into saturation, the power consumption is more.

Transistor-Transistor Logic

TTL logic gate has totem pole arrangement at the output of the circuit. There are two variations in the output of this circuit namely open collector and tri-state. TTL logic family is further classified based on the performance characteristics namely, speed and propagation delay. Accordingly, it is classified as Low Power Schottky TTL, Schottky TTL etc. The subfamilies are then compared based on these key parameters. TTL is the abbreviation of Transistor-Transistor Logic. This logic family is implemented using bipolar junction transistors; NPN or PNP, PN junction diodes and diffused resistors. TTL (Transistor-Transistor Logic) is an incremental improvement over the DTL (Diode Transistor Logic) Family. As the technology progressed, additional improvements were added, so that it became the most widely used type in the design of digital circuits. NAND gate is the basic gate or building block of this logic family.

The Transistor Transistor Logic (TTL) family has heavily dependence on transistors to provide basic operation. Figure shows internal schematic of basic two inputs TTL NAND gate. The basic NAND gate structure consists of three major sections. The input transistor Q1 forms the multi-emitter input stage. Transistor Q2 is a phase splitter. In the output stage, transistor Q3 sits above Q4 forms the totem pole arrangement. Diodes D2 and D3 at input stage protects transistor from negative spikes of voltages at input. Diode D1 makes sure that transistors Q3 and Q4 do not conduct simultaneously.

When both the inputs A and B are HIGH (+5V), both the base-emitter junctions of Q1 are reverse biased and no current flows to the emitters. However, the base collector junction of Q1 is forward biased and the current flows through the base-collector PN junction diode of transistor Q1 into the base of transistor Q2. Transistor Q2 is turned ON to saturation, with the result that transistor Q3 is switched OFF and transistor Q4 is switched ON. This produces a logic LOW at the output. Transistor Q4 is also referred to as the current-sinking or pull-down transistor.

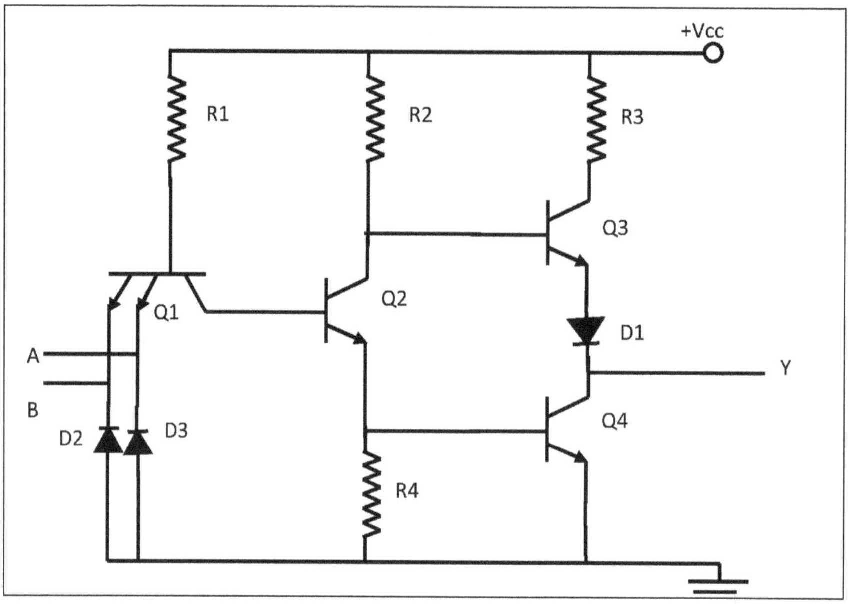

Internal Schematic of TTL NAND gate.

When either of the two inputs A or B or both inputs are LOW state, the base-emitter region of Q1 is forward biased and base collector junction is reverse biased. So the current flows to ground through emitter of transistor Q1. This drives Q2 to cut-off. When Q2 is in the cut-off state, Q3 is driven to conduction and Q4 to cut-off. This produces a logic HIGH output. Transistor Q3, acts as the current-sourcing or pull-up transistor. The totem-pole arrangement at the output has certain distinct advantages. The major advantage of using a totem-pole connection is that, it offers low-output impedance in both the HIGH and LOW output states.

Types of TTL Family

TTL logic families are classified based on the types of output configurations. They are: Open collector output, Totem pole or standard output and Three state (or tri state) output.

Open Collector Output

In the open collector TTL gate, the output stage does not have the active pull-up transistor i.e. Q3 and the output are taken at the collector of transistor Q4. In order to generate required HIGH and LOW states, an external pull-up resistor is connected to +Vcc from the collector of Q4. The Figure shows the internal schematic of a NAND gate with an open collector output.

When transistor Q4 is OFF, the output is pulled to $+V_{cc}$ through the external resistor typically of value 10 kΩ. When Q4 is ON, the transistor is in saturation and the output is nearly at ground potential. The advantage of open collector outputs is that, the outputs of different gates can be wired together, resulting in wire-ANDing of the outputs. This is called wired AND because the AND function is obtained by connecting the outputs of open collector gates together by a simple wire. When any of the NAND gate output goes to a LOW state, then the output of wire AND logic also goes LOW. This is possible because of shorting the transistor to ground through the ON transistor. Similarly, the wire AND output will be HIGH only when all the NAND gate outputs are HIGH.

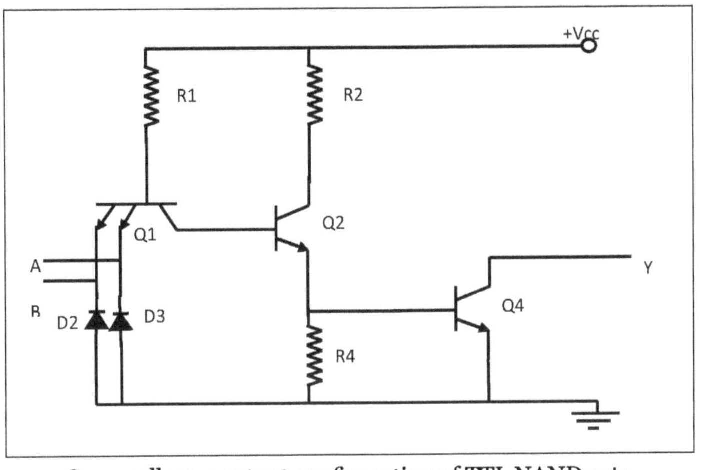

Open collector output configuration of TTL NAND gate.

Tristate Gate

In addition to totem-pole configuration and wire ANDing operation, there is third configuration i.e. Tristate gate, which is also popular. Tristate logic gates have three possible output states, i.e. the logic '1' state, the logic '0' state and a high-impedance state. The high-impedance state is controlled by an external ENABLE input. The ENABLE input decides whether the gate is active or in the high-impedance state.

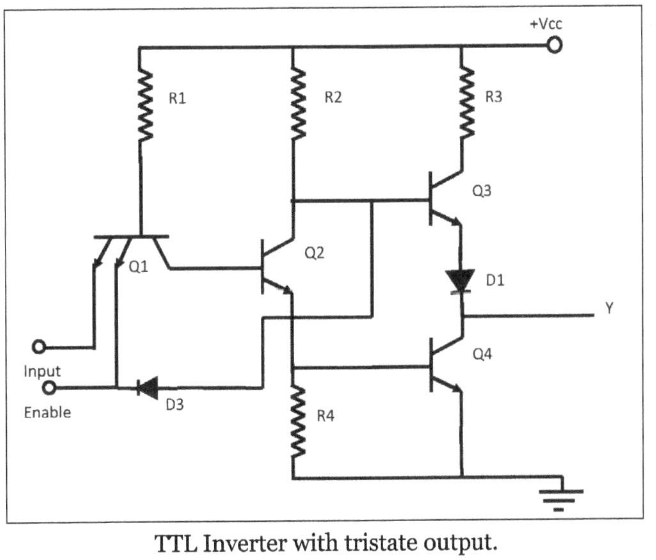

TTL Inverter with tristate output.

Figure shows the circuit configuration for Tristate TTL inverter. The standard TTL NAND circuit is modified so as to act as an inverter with tristate logic. When enable input is active, the output changes the state and can be '0' or '1' depending upon input conditions. One of the main advantages of these gates is that, their inputs and outputs can be connected in parallel to a common bus line.

When the enable input is HIGH, the circuit operates as normal inverter because transistor Q1 and Q2 have no effect and diode D1 is reverse-biased. In this enabled operation, the circuit behaves like an inverter. When the enable input is LOW, diode D1 becomes forward biased. A LOW enable input forces transistor Q2 to cut-off and which in turn forces Q4 to turn-off. Also, when enable is low a forward-biased D1 forces Q3 to cut-off. When both output transistors are in cut-off, the output essentially is an open circuit. Thus the output presents high output impedance state.

TTL Subfamilies

The speed-power product is an important parameter for comparing the basic gates. It is defined as the product of propagation delay and power dissipation measured in picojoules (pJ). Value of this parameter indicates that, a given propagation delay can be achieved without excessive power dissipation. This product is also referred to as figure of merit for the logic family. Depending upon the value of figure of merit, a minor variation in the standard TTL family is implemented using components with either a less power dissipation or a high speed. These are as follows:

- Standard TTL: The standard TTL was the first gate in TTL family. The gate was constructed using different resistor values to produce gates with lower dissipation or high speed. The propagation delay of a saturated logic family depends largely on two factors: storage time and RC time constants. Reducing storage time reduces the RC time constant and decreases the propagation delay.
- Low Power TTL(L TTL): The low power TTL gate has resistor values higher than those in standard TTL so as to reduce power dissipation and increased propagation delay.
- High Speed TTL(H TTL): In high speed TTL the resistor values are lowered to reduce the propagation delay but power dissipation is increased.
- Schottky TTL(S TTL): The schottky TTL removes the storage time of transistors by preventing them from going into saturation. In this way, the speed of operation increases without excessive increase in power dissipation from Schottky TTL. Basic TTL family has a speed limitation due to turn off delays of transistors which are caused by transition from saturation to cutoff. This limitation can be removed by replacing the transistors of TTL gate by Schottky transistors. These transistors are prevented from entering the saturation and hence there is saving in turn-off time. The propagation delay of shottky TTL is of the order of 2 ns. In comparison, the propagation delay for standard TTL is 10 ns. The Schottky TTL therefore is a non-saturating bipolar logic.
- Low Power Schottky TTL(LS TTL): The low power schottky TTL sacrifices speed for areduced power dissipation. It has the same propagation delay as standard TTL, but the power dissipation is reduced to a 1/5th that of standard TTL. It is the most popular version of in new digital circuit system design.

TTL versions are available in SSI, MSI and LSI packages. The TTL versions do not vary functionally; they vary in terms of the values of resistors used and the type of transistors that the basic gate uses. These subfamilies are compared on the basis of figure merit as shown in the Table.

Table: Comparison of TTL subfamilies.

Name	Abbreviation	Propagation delay (ns)	Power Dissipa-tion(mW)	Figure of merit Speed power product(PJ)
Standard TTL	TTL	10	10	100
Low power TTL	LTTL	33	1	33
High speed TTL	HTTL	6	22	132
Schottky TTL	STTL	3	19	57
Low power schottky TTL	LSTTL	9.5	2	19

Characteristics of TTL Family

TTL 5400/7400 series is the most popular and commonly used series in digital ICs. 7400 devices are used for commercial applications whereas the 5400 devices are used for military applications. The only difference in two series is in the temperature range and power supply range. The temperature range for 7400 series is 0 °C to 70 °C, and that for 5400 series is -55 °C to 125 °C. The supply voltage range is 5±0.25V for 7400 and is 5±0.5V for 5400 series. There are five categories of TTL ICs. These are as listed in Table.

Table: TTL subfamilies.

Category	Prefix	Examples
Standard TTL	74-	7402, 74193
High power TTL	74H-	74H02, 74H193
Low power TTL	74L-	74L02, 74L193
Schottky TTL	74S-	74S02, 74S193
Low power Schottky TTL	74LS-	74LS02, 74LS193

It is observed that, for all subfamilies input and output voltage and current level requirements are almost same. However, one can distinguish them on the basis of total propagation delay. In Low power, TTL it is about 30ns whereas for schottky TTL is very small of the order of 2.5ns. In low power schottky TTL it is about 15 ns. It is a trade- off between the LTTL and STTL.

Table: Electrical characteristics of TTL subfamilies.

Parameter	5400/7400	54H00/74H00	54L00/74L00	54S00/74S00	54LS00/74LS00	Units
VIH	2	2	2	2	2	Volts
VIL 54 74	0.8 0.8	0.8 0.8	0.7 0.7	0.8 0.8	0.7 0.8	Volts
VOH 54 74	2.4 2.4	2.4 2.4	2.4 2.4	2.5 2.7	2.5 2.7	Volts
VOL 54 74	0.4 0.4	0.4 0.4	0.3 0.4	0.5 0.5	0.4 0.5	Volts
IIH	40	50	10	50	20	µA
IIL	-1.6	-2.0	-0.18	-2.0	-0.36	mA

IOH	16	20	2	20	4	μA
IOL 54 74	16 16	20 20	2 3.6	20 20	4 8	mA
ICC(1)	8	16.8	0.8	16	1.6	mA
ICC(0)	22	40	2.04	36	4.4	mA
tpHL	15	10	60	5	15	ns
tpLH	22	10	60	4.5	15	ns

Fan-out capabilities of different subfamilies, are dependent on the output source current capabilities. It is observed from the table that, if the source device is S TTL and output device is L TTL maximum number of gates can be connected.

Table: Summary to TTL fan out capabilities.

Source TTL Device	5400/7400	54H00/74H00	54L00/74L00	54S00/74S00	54LS00/74LS00
5400/7400	10	8	40	8	20
54H00/74H00	12	10	50	10	25
54L00/74L00	2	1	20	1	10
54S00/74S00	12	10	100	10	50
54LS00/74LS00	5	4	40	4	20

Handling Unused TTL Inputs

The floating input of TTL family devices behaves as if a logic HIGH has been applied to the input. Such a behaviour is explained from the input circuit of a TTL device as shown in Figure. When the input is HIGH, the input emitter-base junction is reverse biased, and the current that flows into the input is the reverse-biased diode leakage current. The input diode will be reverse biased, even when the input terminal is left unconnected or floating. This implies that a floating input behaves as if there were a logic HIGH applied to it. As an initial thought, we may tend to believe that it should not make any difference if we leave the unused inputs of NAND and AND gates as floating, as a logic HIGH- like behaviour of the floating input makes no difference to the logical behaviour of the gate, as shown in Figures (a) and (b).

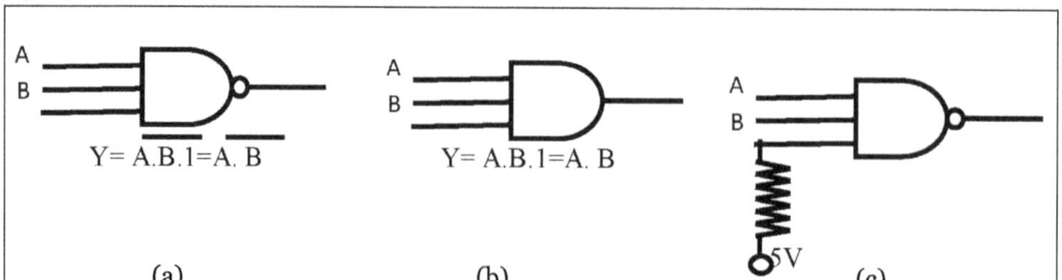

Handling unused outputs of AND and NAND gates.

In spite of this, it is strongly recommended that the unused inputs of AND and NAND gates be connected to a logic HIGH input because, floating input behaves as an antenna and may pick up stray noise and interference signals, thus causing the gate to function improperly. 1 kΩ resistance is connected to protect the input from any current spikes caused by any spikes on the power supply line. More than one unused input (up to 50) can share the same 1 kΩ resistance, if needed.

In the case of OR shown in Figures (a) and (c) below and NOR gates, unused inputs are connected to ground (logic LOW), as shown in Figure (b), for obvious reasons. A floating input or an input tied to logic HIGH in this case produces a permanent logic HIGH (for an OR gate) and LOW (for a NOR gate) at the output. An alternative solution is shown in figures where the unused input is tied to one of the used inputs. This solution works well for all gates, but one has to be conscious of the fact that the fanout capability of the output driving the tied inputs is not exceeded.

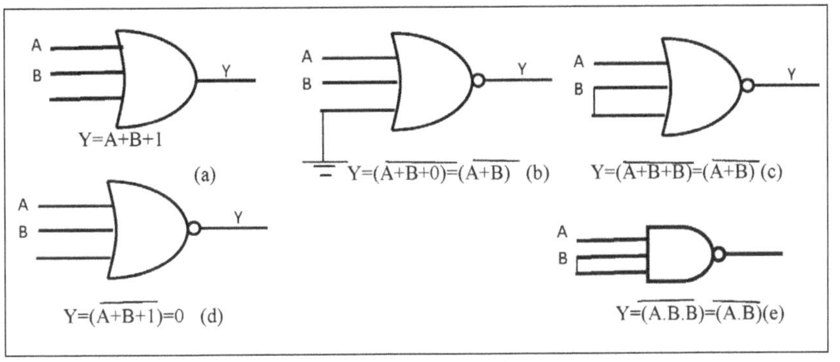

Unused inputs of NAND and NOR gates.

Current Transients and Power Supply Decoupling

TTL family devices are prone to occurrence of narrow-width current spikes on the power supply line. Current transients are produced, when the totem-pole output stage of the device undergoes a transition from a logic LOW to a logic HIGH state. The problem becomes severe when in a digital circuit a large number of gates are likely to switch states at the same time. These current spikes produce voltage spikes due to any stray inductance present on the line. On account of the large rate of change in current in the current spike, even a small value of stray inductance produces voltage spikes large enough to adversely affect the circuit performance.

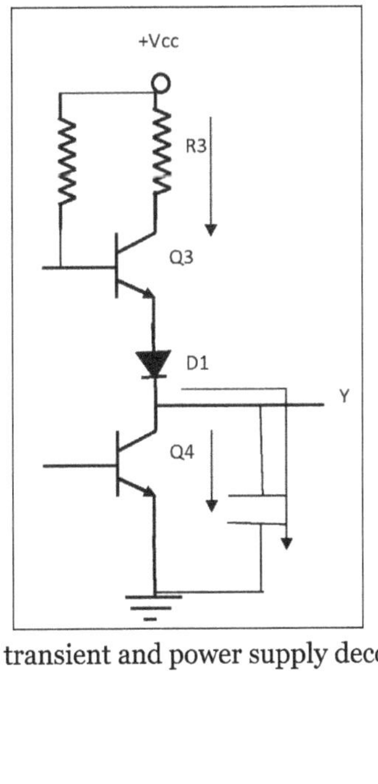

Current transient and power supply decoupling.

Figure illustrates the phenomenon. When the output changes from LOW to HIGH, there is a small time interval during which both the transistors are conducting; because the pull-up transistor Q3 has switched on and the pull-down transistor Q4 has not yet come out of saturation. During this time interval, there is an increase in current drawn from the supply; ICCL experiences a positive spike before it settles down to a usually lower ICCH. The presence of any stray capacitance C across the output owing to any stray wiring capacitance or capacitance loading of the circuit being fed also adds to the problem. The problem of voltage spikes on the power supply line is usually overcome by connecting small-value, low-inductance, high-frequency capacitors between VCC terminal and ground. It is standard practice to use a 0.01 or 0.1 _F ceramic capacitor from VCC to ground. This capacitor is also known by the name of power supply decoupling capacitor, and it is recommended to use a separate capacitor for each IC. A decoupling capacitor is connected as close to the VCC terminal as possible, and its leads are kept to a bare minimum to minimize lead inductance. In addition, a single relatively large value capacitor in the range of 1–22 μF is also connected between VCC and ground on each circuit card to take care of any low-frequency voltage fluctuations in the power supply line.

MOS Logic Family

The MOS logic family is based on Metal Oxide Semiconductor Field Effect Transistor (MOSFET) devices. As compared to bipolar logic families, the MOS families are simpler and low power devices. A field effect transistor is made by growing a very thin layer of SiO_2 over a semiconductor material. A metal such as aluminum is deposited over the dielectric layer of SiO_2 to obtain MOS device.

Enhancement and Depletion Mode

There are two types of MOSFETs; Enhancement type MOSFET and Depletion type MOSFET. Just Similar to junction field effect transistors, there are n-channel as well as p-channel MOSFETs. N-channel devices are more popular than p-channel devices because of their higher speed. Basic principle of operation of p-channel and n-channel MOSFETs is the same. P-channel MOSFETs are considered as obsolete for independent or individual devices but are found to be useful in complementary MOS (CMOS) technology.

Enhancement MOSFET

Enhancement type MOSFETs are the popular switching devices in most MOS families. The basic structure of n-channel enhancement type MOSFET is shown in Figure. Source and drain are formed by two n-type regions which are diffused into a p-type substrate. A thin layer of silicon dioxide is grown over this and by masking, etching and metallization processes, metallic contacts are taken from the source and the drain. A metal gate is formed above the dielectric layer.

Initially, without bias condition, has no channel between the source and the drain. A positive voltage greater than threshold voltage V_T is applied at the gate. Due to this field from the gate, the electronics (minority charge carriers) from the substrate are attracted towards the gate between the source and the drain regions. In the process, they form an n channel by making the region between the source and the drain n type. This channel makes it possible for the electrons to flow from the source to the drain, when a positive voltage is applied at the drain. Since the application of a positive voltage at the gate enhances the channel width, this device is referred to as an enhancement mode device or an enhancement MOSFET. An increase in the drain voltage increases the drain current thereby producing resistor type operation for small voltages similar to JFET. A pinch-off occurs, when the drain to source voltage is sufficiently large, which reduces the field near the drain to zero and makes the drain current relatively constant. These devices are normally OFF when Vgs=0 and can be turned ON by making Vgs greater than source voltage for n-MOS or lesser than the source voltage for p-MOS.

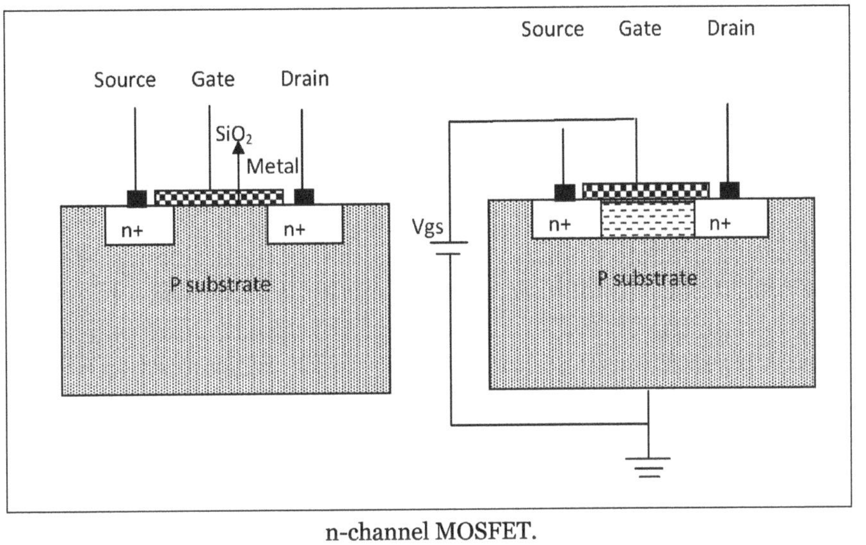

n-channel MOSFET.

Depletion MOSFET

Depletion MOS device is fabricated by diffusing n type of impurity between the two n-type regions, which acts as a channel between the source and the drain. In this device, current flows between drain and source even in the absence of a positive voltage at the gate. This causes depletion of the channel and hence it is called depletion MOSFET. The circuit symbol for n channel and p channel MOSFETs are shown in Figure. The depletion-mode MOSFETs are normally ON at zero gate-source voltage. Such devices are used as load resistors in logic circuits.

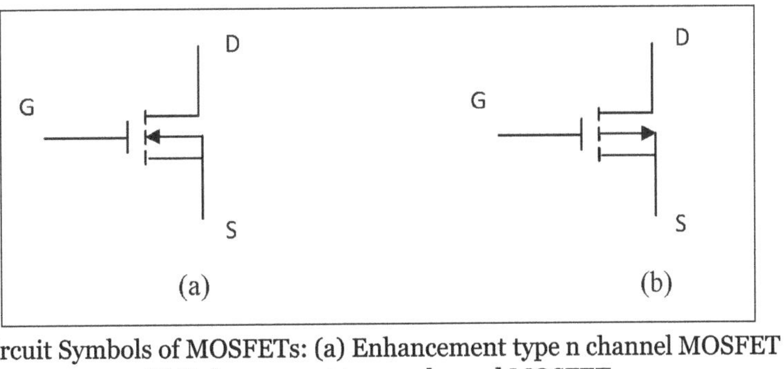

Circuit Symbols of MOSFETs: (a) Enhancement type n channel MOSFET
(b) Enhancement type p channel MOSFET.

MOS Logic family can be classified into three categories. They are:

- NMOS logic family: Built with N-channel Metal oxide semiconductor FET (MOSFET).

- PMOS logic family: Built with P-channel Metal oxide semiconductor FET (MOSFET).
- CMOS logic family: Built with both N-channel and P-channel Metal oxide semiconductor FET (MOSFET).

Off all the MOS families, NMOS and CMOS are used in making an integrated circuit. It is because of its advantages.

NMOS Logic Circuits

In the N-channel MOS family, current conduction is because of the electrons. The negatively charged electrons are fast-moving than holes, which is positively charged. So the speed of operation of NMOS is faster than PMOS.

NMOS Inverter

The NMOS inverter circuit has two N-channel MOSFET devices. Among the two MOSFETs, Q_1 acts as the load MOSFET, and Q_2 acts as a switching MOSFET. Since the gate is always connected to the supply $+V_{DD}$, the MOSFET Q_1 is always ON. So, the internal resistance of Q_1 acts as the load resistance R_L. Compared to both MOSFETs, Q_1 is designed to have more resistance than Q_2.

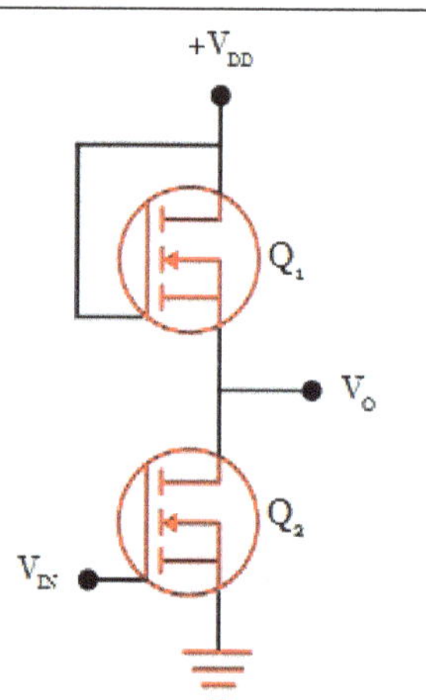

V_{IN}	Q_2	V_O
Logic 0 (LOW)	OFF	Logic 1 (HIGH)
Logic 1 (HIGH)	ON	Logic 0 (LOW)

When zero voltage(LOW input) is given at the gate terminal of Q_2, it is turned OFF. It will make the output V_O to be HIGH, which is the supply voltage. When HIGH input is given at the gate terminal, the MOSFET Q_2 will be turned ON, which will make the output V_O be at LOW value.

NMOS NAND Gate

The figure shown below is the circuit of the 2-input NMOS NAND gate. It consists of three N-channel MOSFETs, in which Q_1 acts as the load resistance, whereas Q_2 and Q_3 act as the switching MOSFETs. The two inputs A and B are given to MOSFET Q_2 and Q_3 respectively.

When both inputs A and B are given LOW input, both the MOSFETs are turned OFF, which makes the output V_O as HIGH. When either A or B is LOW, the MOSFET with low input will be turned

OFF, thus making the output to be HIGH. If HIGH input is given to both the input terminals, it will turn ON the MOSFETs Q_2 and Q_3. Therefore, the current due to the supply voltage V_{DD} will flow towards the ground making the output as LOW.

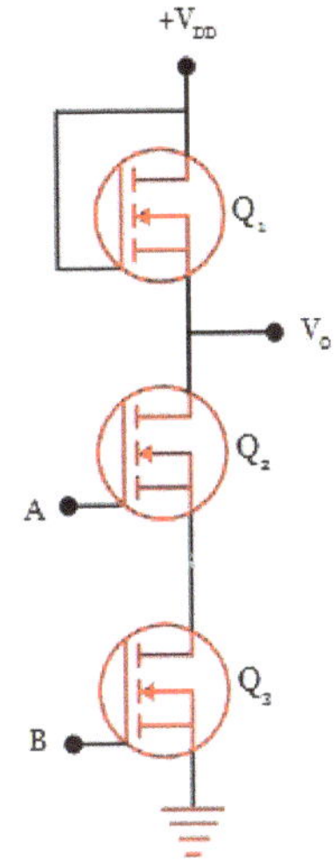

A	B	Q_2	Q_3	V_o
0	0	OFF	OFF	1(HIGH)
0	1	OFF	ON	1(HIGH)
1	0	ON	OFF	1(HIGH)
1	1	ON	ON	0 (LOW)

NMOS NOR Gate

The following circuit shows the circuit of the 2-input NMOS NOR gate. It has 3 N-channel MOSFETs, in which Q_1 acts as the load resistance, MOSFETs Q_2 and Q_3 act as switching devices. If Low input is given at both the inputs, it will turn OFF both the MOSFETs Q_2 and Q_3, thereby making the output to be HIGH.

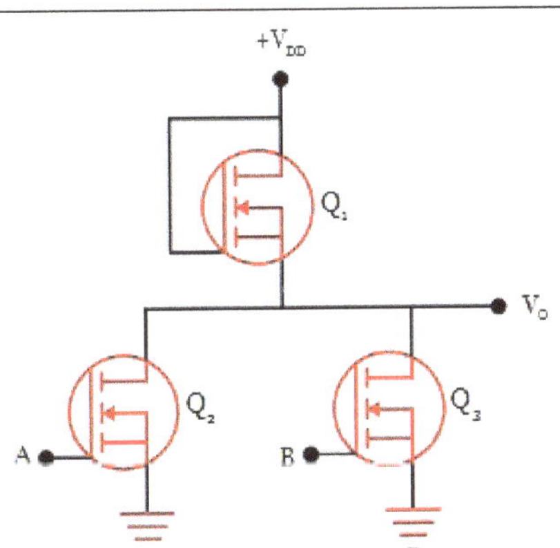

A	B	Q_2	Q_3	V_o
0	0	OFF	OFF	1(HIGH)
0	1	OFF	ON	0 (LOW)
1	0	ON	OFF	0 (LOW)
1	1	ON	ON	0 (LOW)

2-input NMOS NOR gate.

If any of the input, either A or B is given HIGH input, the corresponding MOSFET with HIGH input will turn ON making the output to be LOW. Similarly, if both the inputs are HIGH, it will turn ON both the MOSFETs, which will make the output to be LOW.

PMOS Logic Circuits

The digital logic family can be built with P-channel MOSFETs and those circuits are known as PMOS logic circuits. The operation of PMOS is similar to the NMOS circuits, except that the mode of conduction is different. For a P-channel MOSFET, a negative voltage is to be given at the gate terminal to create a channel. Thus negative voltage (LOW voltage) is enough to turn ON the PMOS devices.

CMOS Logic Circuit Family

CMOS(Complementary MOS) logic family uses both N-channel and P-channel MOSFET devices. CMOS has greater complexity than PMOS and NMOS. However, the speed of operation is high and power dissipation is less in CMOS. CMOS also has more fan-out and better noise margin.

CMOS Inverter

In CMOS inverter, both the n-channel and p-channel devices are connected in series. The source terminal of the P-channel device is connected to source voltage $+V_{DD}$. The source terminal of the N-channel device is connected to the ground. The gate of both the devices are connected together and a common input is given to both the MOSFET device. The drain terminals are connected together as a common output.

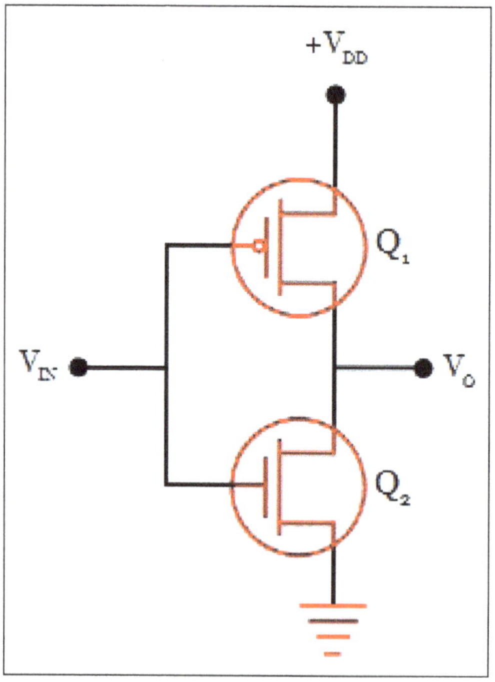

For a HIGH input at VIN, the P-channel MOSFET(Q_1) gets turned OFF, but the N-channel device(Q_2) will be turned ON. This will drive the output V_o to be at Logic LOW. If LOW input is given at the input terminal VIN, it will turn ON on Q_1 and turn OFF Q_2, making the output to be HIGH.

CMOS NAND Gate

The circuit shown below shows the circuit of the 2-input CMOS NAND gate. It has two p-channel MOSFETs (Q_1, Q_2) and two n-channel MOSFETs (Q_3 and Q_4). A and B are two inputs. The input A is given to the gate terminal of Q_1 and Q_3. The input B is given to the gate terminal of Q_2 and Q_4. The output is obtained from the terminal V_O.

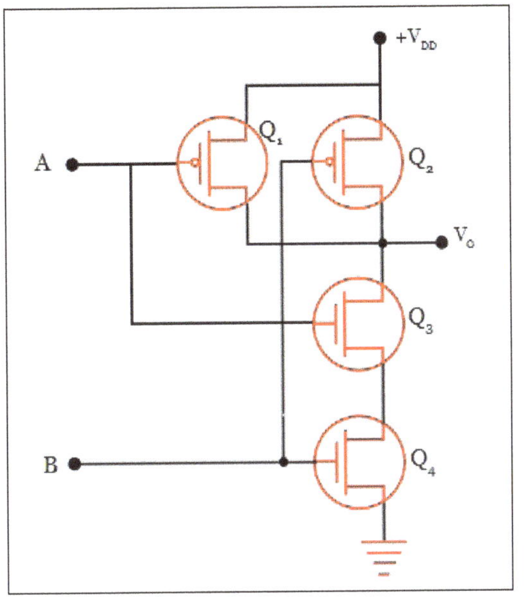

When both the inputs are given LOW input, it will turn ON Q_1, Q_2 and turn OFF the MOSFETs Q_3 and Q_4. The output terminal is connected to the supply voltage VDD and the output will be HIGH. It is shown in the below figure (a). When either one of the input is high, for e.g. Let us consider A is given HIGH input and B is given LOW input. In this case, MOSFETs Q_1 and Q_4 get turned ON, whereas Q_2 and Q_3 are turned OFF as shown in the figure (b) below. This will make a path for the supply voltage to be connected to the load, making the output to be HIGH. If both the inputs A and B are HIGH inputs, which make the MOSFETs Q_3, Q_4 to be turned ON and Q_1, Q_2 to be turned OFF. Thus the output is connected to the ground alone as shown below(c). Thereby, the output will be at LOW value.

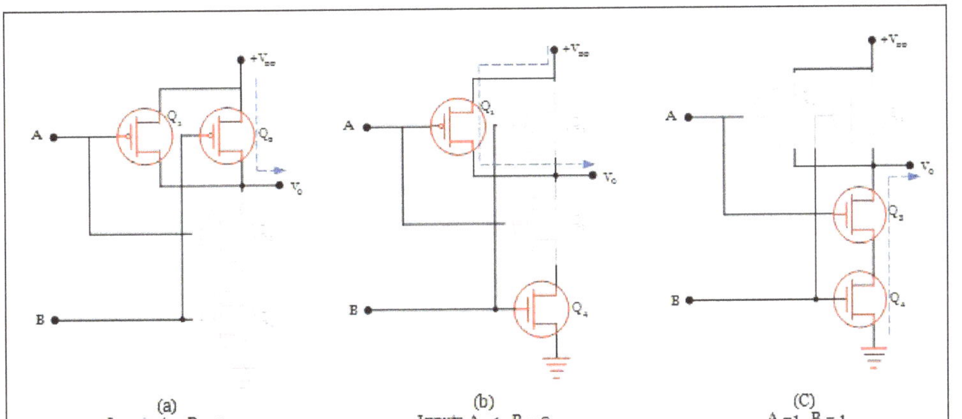

CMOS NOR Gate

Similar to other logic family, CMOS NOR gate circuit also has two NMOS and two PMOS devices and the input and output are connected as shown in the below figure. The operation of 2-input CMOS NOR gate is shown in the figure.

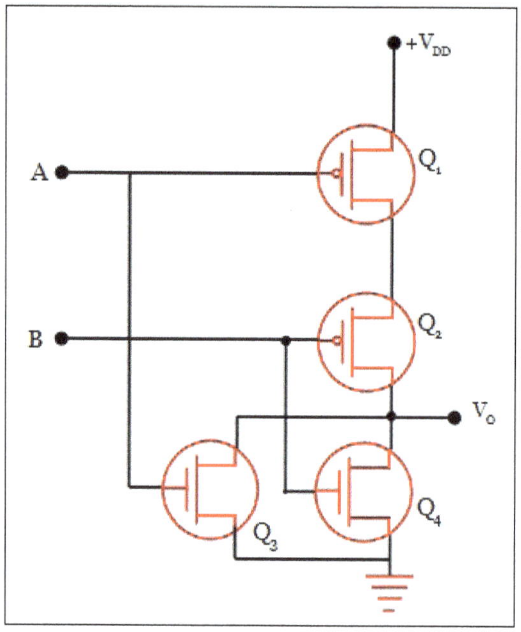

For the LOW inputs at A and B, PMOS devices Q_1 and Q_2 will conduct, making the output to be at logic HIGH. When any one of the input is LOW, it will produce a LOW output as shown in the below figure (b). If both A and B are given HIGH input, it will turn ON the PMOS devices Q_3 and Q_4, making the output voltage to be logic LOW. It is shown in figure (c) below:

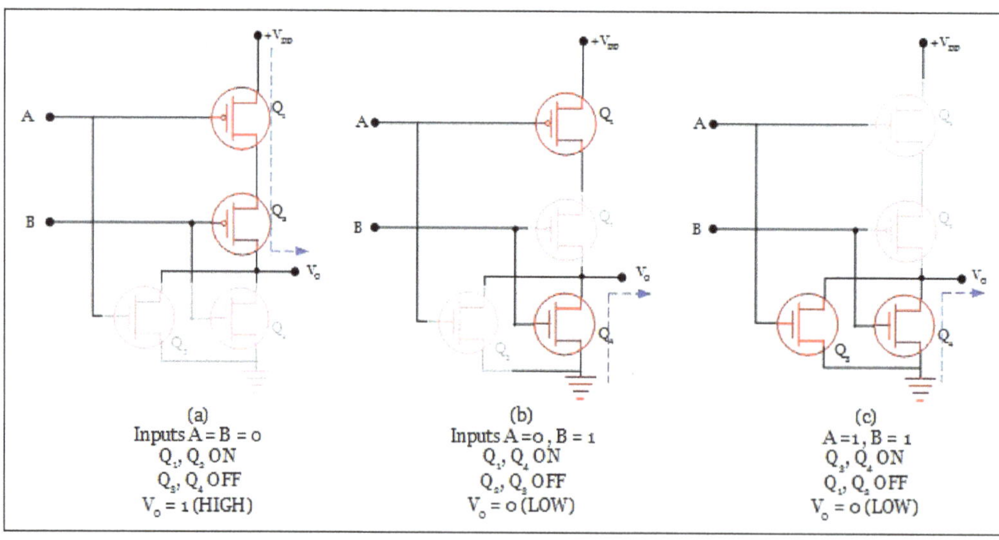

CMOS Working Principle

In CMOS technology, both N-type and P-type transistors are used to design logic functions. The same signal which turns ON a transistor of one type is used to turn OFF a transistor of the other type. This characteristic allows the design of logic devices using only simple switches, without the need for a pull-up resistor.

In CMOS logic gates a collection of n-type MOSFETs is arranged in a pull-down network between the output and the low voltage power supply rail (Vss or quite often ground). Instead of the load resistor of NMOS logic gates, CMOS logic gates have a collection of p-type MOSFETs in a pull-up network between the output and the higher-voltage rail (often named V_{dd}).

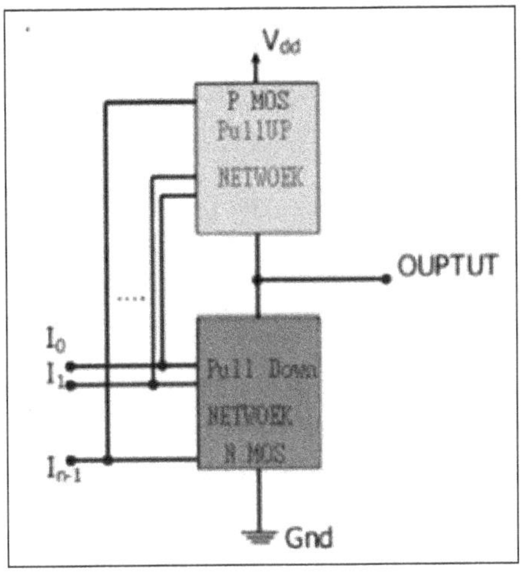

CMOS using Pull Up anf Pull Down.

Thus, if both a p-type and n-type transistor have their gates connected to the same input, the p-type MOSFET will be ON when the n-type MOSFET is OFF, and vice-versa. The networks are arranged such that one is ON and the other OFF for any input pattern. CMOS offers relatively high speed, low power dissipation, high noise margins in both states, and will operate over a wide range of source and input voltages (provided the source voltage is fixed).

CMOS Fabrication

For less power dissipation requirement CMOS technology is used for implementing transistors. If we require a faster circuit then transistors are implemented over IC using BJT. Fabrication of CMOS transistors as IC's can be done in three different methods:

- The N-well/P-well technology, where n-type diffusion is done over a p-type substrate or p-type diffusion is done over n-type substrate respectively.
- The Twin well technology, where NMOS and PMOS transistor are developed over the wafer by simultaneous diffusion over an epitaxial growth base, rather than a substrate.
- The silicon On Insulator process, where rather than using silicon as the substrate an insulator material is used to improve speed and latch-up susceptibility.

N-Well/P-Well Technology

CMOS can be obtained by integrating both NMOS and PMOS transistors over the same silicon wafer. In N–well technology an n-type well is diffused on a p-type substrate whereas in P- well it is vice- verse.

CMOS Fabrication Steps

The CMOS fabrication process flow is conducted using twenty basic fabrication steps while manufactured using N- well/P-well technology.

Making of CMOS using N Well

- Step 1 - First we choose a substrate as a base for fabrication. For N-well, a P-type silicon substrate is selected.

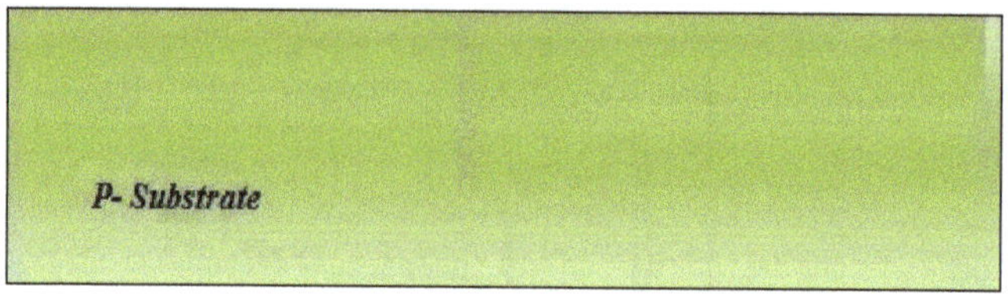

Substrate.

- Step 2 - Oxidation: The selective diffusion of n-type impurities is accomplished using SiO_2 as a barrier which protects portions of the wafer against contamination of the substrate. SiO_2 is laid out by oxidation process done exposing the substrate to high-quality oxygen and hydrogen in an oxidation chamber at approximately 1000 °c.

Oxidation.

- Step 3 - Growing of Photoresist: At this stage to permit the selective etching, the SiO_2 layer is subjected to the photolithography process. In this process, the wafer is coated with a uniform film of a photosensitive emulsion.

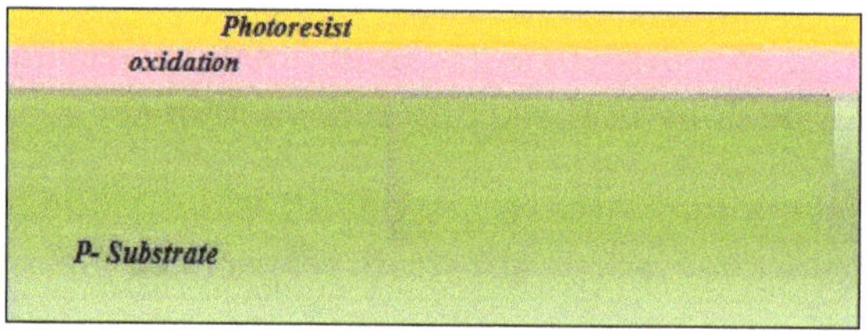

Growing of Photoresist.

- Step 4 - Masking: This step is the continuation of the photolithography process. In this step, a desired pattern of openness is made using a stencil. This stencil is used as a mask over the photoresist. The substrate is now exposed to UV rays the photoresist present under the exposed regions of mask gets polymerized.

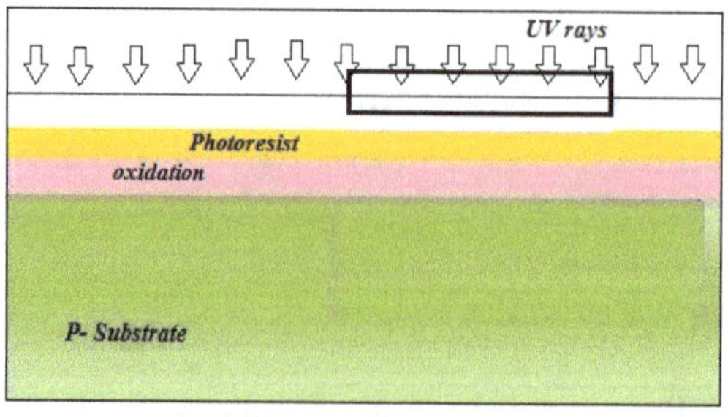

Masking of Photoresist.

- Step 5 - Removal of Unexposed Photoresist: The mask is removed and the unexposed region of photoresist is dissolved by developing wafer using a chemical such as Trichloroethylene.

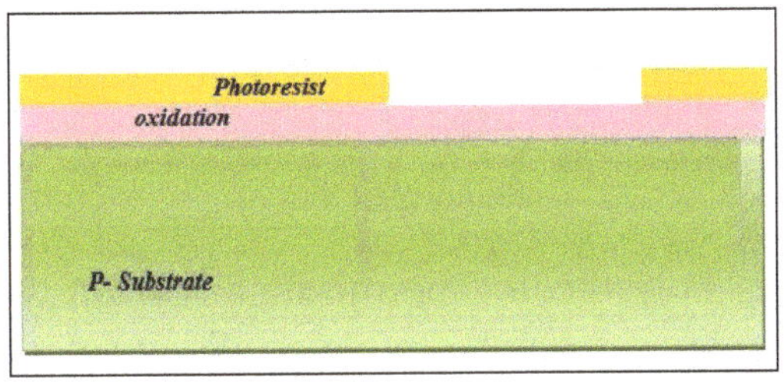

Removal of Photoresist.

- Step 6 - Etching: The wafer is immersed in an etching solution of hydrofluoric acid, which removes the oxide from the areas through which dopants are to be diffused.

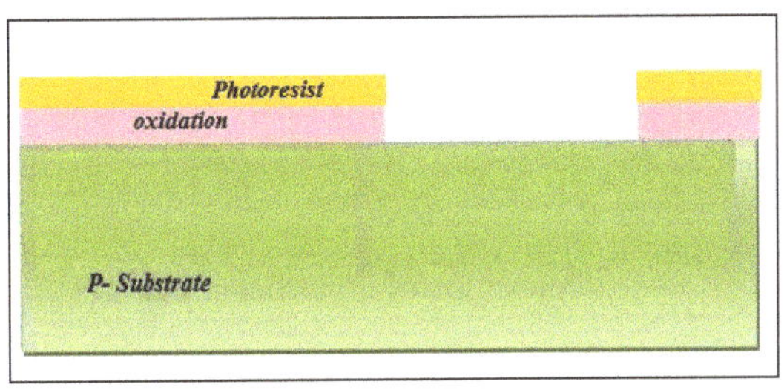

Etching of SiO_2.

- Step 7 - Removal of Whole Photoresist Layer: During the etching process, those portions of SiO_2 which are protected by the photoresist layer are not affected. The photoresist mask is now stripped off with a chemical solvent (hot H_2SO_4).

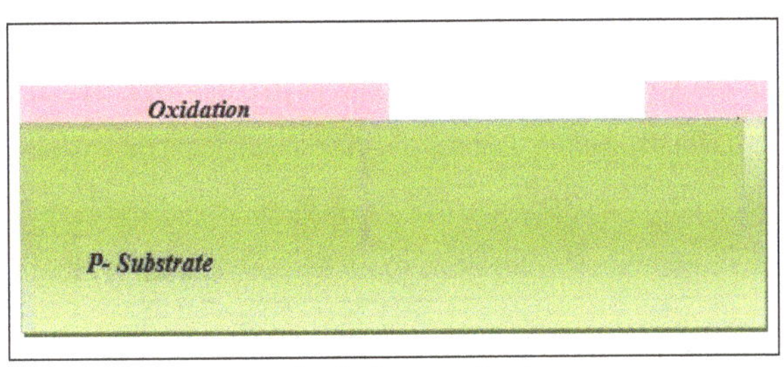

Removal of Photoresist Layer.

- Step 8 - Formation of N-well: The n-type impurities are diffused into the p-type substrate through the exposed region thus forming an N-well.

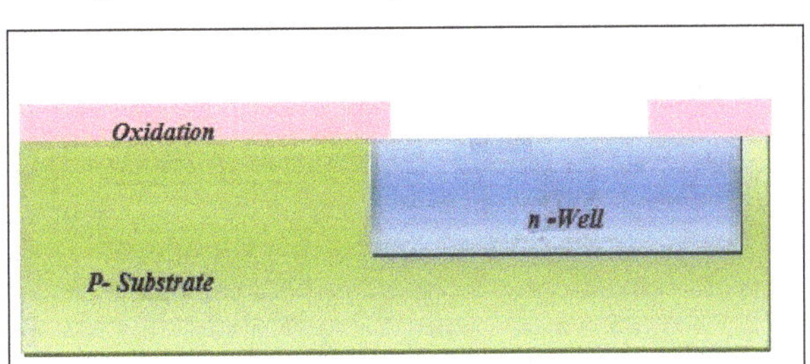

Formation of N-well.

- Step 9 - Removal of SiO_2: The layer of SiO_2 is now removed by using hydrofluoric acid.

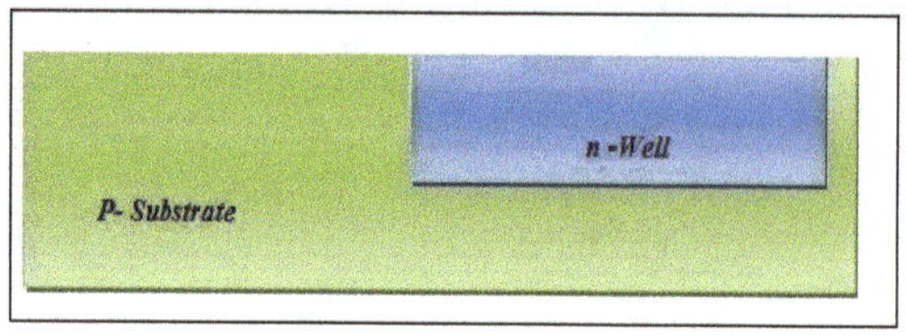

Removal of SiO_2.

- Step 10 - Deposition of Polysilicon: The misalignment of the gate of a CMOS transistor would lead to the unwanted capacitance which could harm circuit. So to prevent this "Self-aligned gate process" is preferred where gate regions are formed before the formation of source and drain using ion implantation.

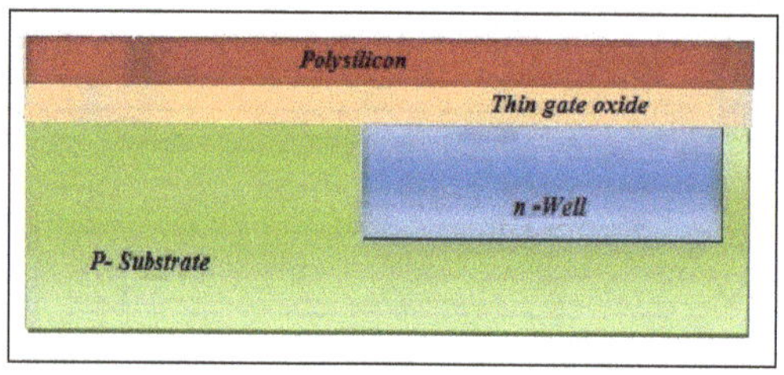

Deposition of Polysilicon.

Polysilicon is used for formation of the gate because it can withstand the high temperature greater than 8000 °c when a wafer is subjected to annealing methods for formation of source and drain. Polysilicon is deposited by using Chemical Deposition Process over a thin layer of gate oxide. This thin gate oxide under the Polysilicon layer prevents further doping under the gate region.

- Step 11 - Formation of Gate Region: Except the two regions required for formation of the gate for NMOS and PMOS transistors the remaining portion of Polysilicon is stripped off.

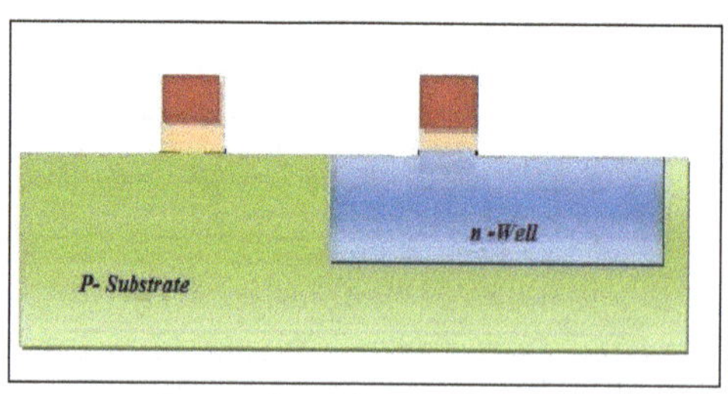

Formation of Gate Region.

- Step 12 - Oxidation Process: An oxidation layer is deposited over the wafer which acts as a shield for further diffusion and metallization processes.

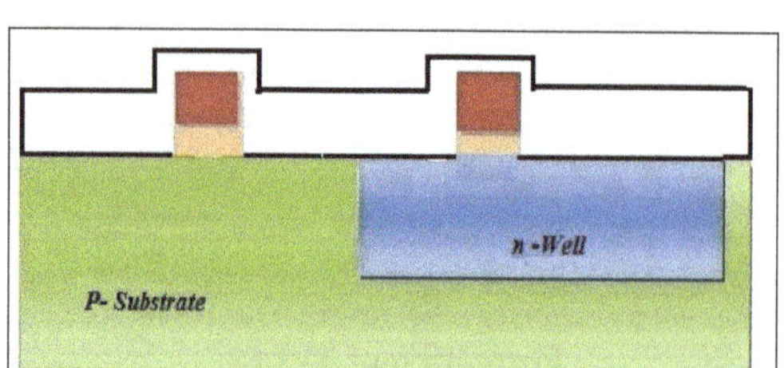

Oxidation Process.

- Step 13 - Masking and Diffusion: For making regions for diffusion of n-type impurities using masking process small gaps are made. Using diffusion process three n+ regions are developed for the formation of terminals of NMOS.

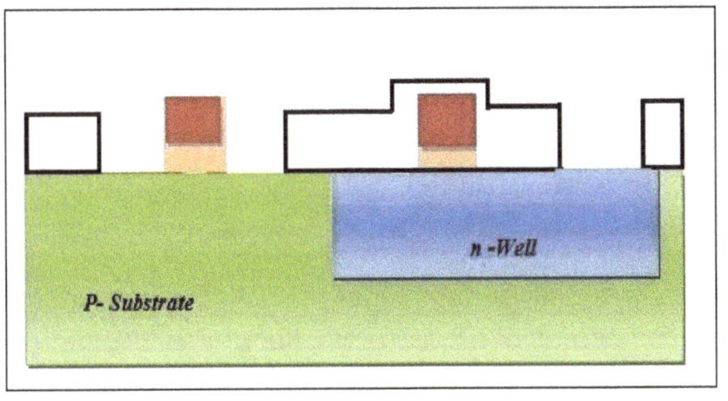

Masking.

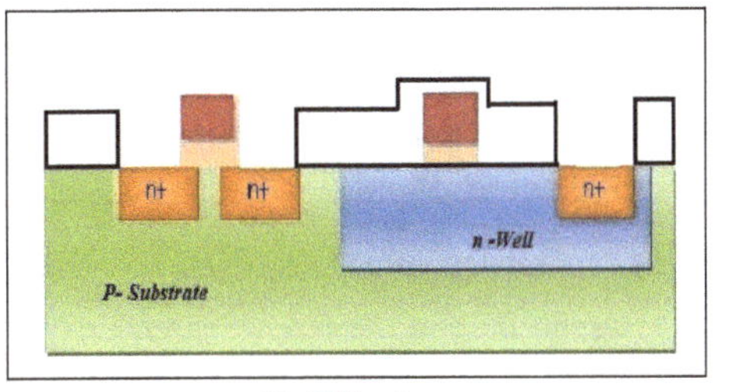

N-diffusion.

- Step 14 - Removal of Oxide: The oxide layer is stripped off.

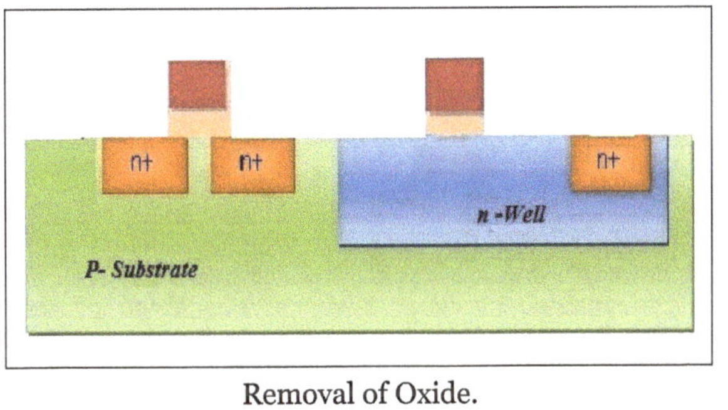

Removal of Oxide.

- Step 15 - P-type Diffusion: Similar to the n-type diffusion for forming the terminals of PMOS p-type diffusion are carried out.

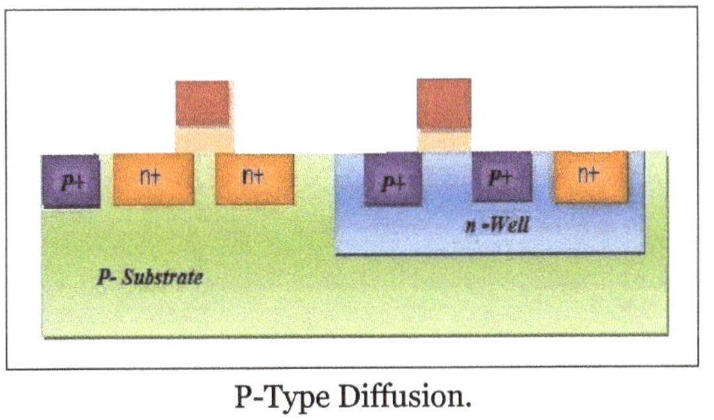

P-Type Diffusion.

- Step 16 - Laying of Thick Field oxide: Before forming the metal terminals a thick field oxide is laid out to form a protective layer for the regions of the wafer where no terminals are required.

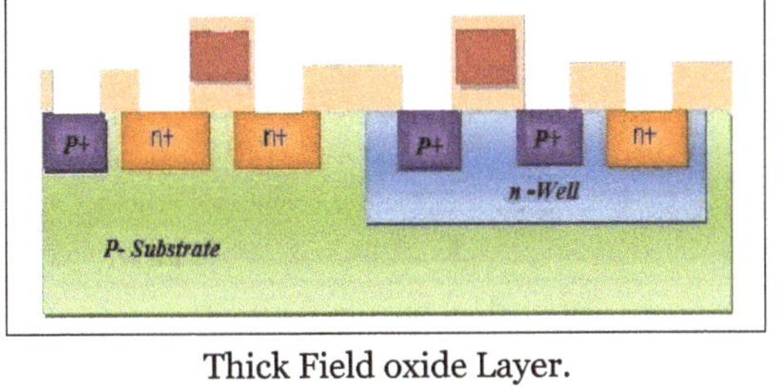

Thick Field oxide Layer.

- Step 17 - Metallization: This step is used for the formation of metal terminals which can provide interconnections. Aluminum is spread on the whole wafer.

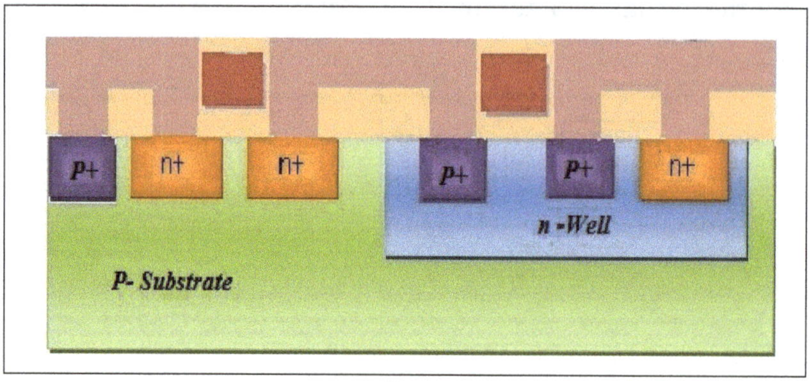

Metallization.

- Step 18 - Removal of Excess Metal: The excess metal is removed from the wafer.
- Step 19 - Formation of Terminals: In the gaps formed after removal of excess metal terminals are formed for the interconnections.

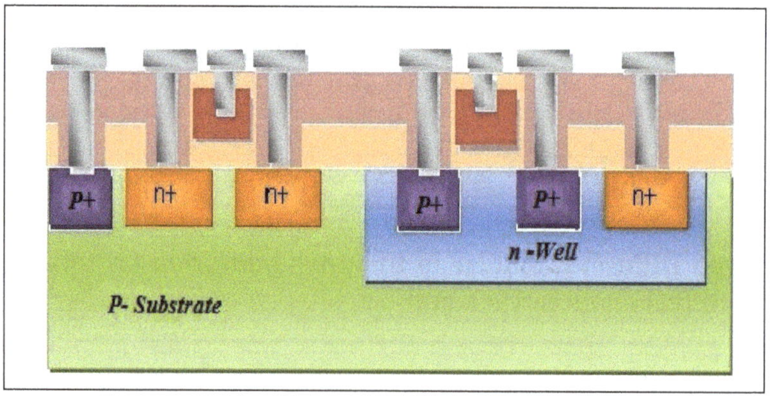

Formation of Terminals.

- Step 20 - Assigning the Terminal Names: Names are assigned to the terminals of NMOS and PMOS transistors.

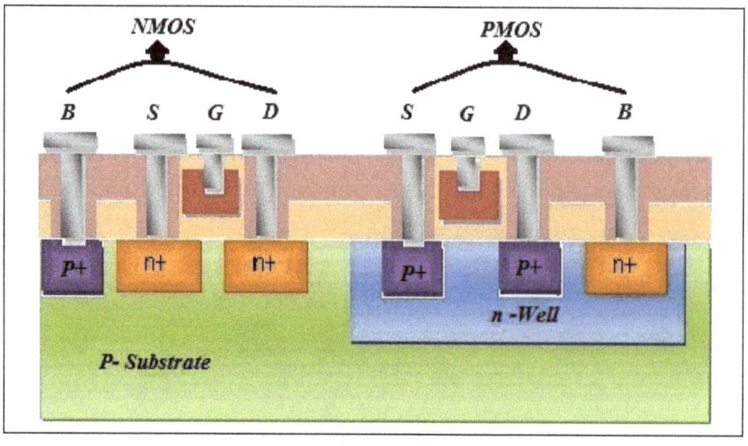

Assigning Terminal names.

Making of CMOS using P Well Technology

The p-well process is similar to N well process except that here n-type substrate is used and p-type diffusions are carried out. For simplicity usually, N well process is preferred.

Twin Tube Fabrication of CMOS

Using Twin-tube process one can control the gain of P and N-type devices. Various steps involved in the fabrication of CMOS using Twin-tube method are as follows:

- A lightly doped n or p-type substrate is taken and the epitaxial layer is used. Epitaxial layer protects the latch-up problem in the chip.
- The high purity silicon layers with measured thickness and exact dopant concentration are grown.
- Formation of tubes for P and N well.
- Thin oxide construction for protection from contamination during diffusion processes.
- Source and drain are formed using ion implantation methods.
- Cuts are made for making portions for metal contacts.
- Metallization is done for drawing metal contacts.

CMOS IC Layout

The upper view of a CMOS fabrication and layout is given. Here various metal contacts and N well diffusions can be viewed clearly. Thus, this is all about CMOS fabrication techniques. Let us consider a 1-in-square wafer divided into 400 chips of surface area 50 mil by 50 mils. It takes an area of 50 mil2 to fabricate a transistor. Hence each IC contains 2 transistors thus there are 2 x 400 = 800 transistors built on each wafer. If 10 wafers are processed each batch then 8000 transistors can be manufactured simultaneously.

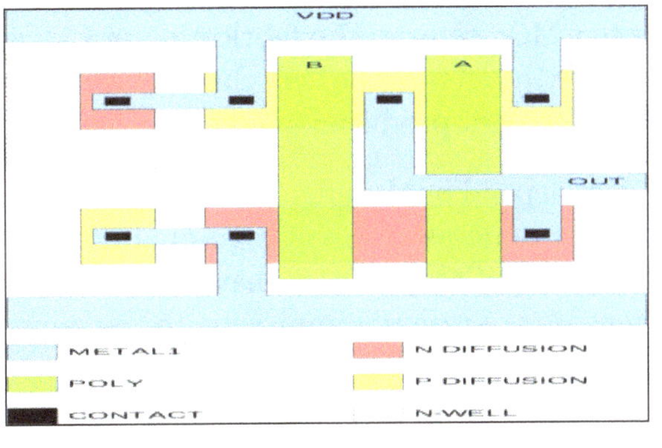

CMOS Characteristics

The most important characteristics of CMOS are low static power utilization, huge noise immunity. When the single transistor from the pair of MOSFET transistor is switched OFF then the series combination uses significant power throughout switching among the two stated like ON & OFF. As a result, these devices do not generate waste heat as compared with other types of logic circuits such as TTL or NMOS logic, which usually use some standing current even they don't change their state.

These CMOS characteristics will allow for integrating logic functions with high density on an integrated circuit. Because of this, CMOS has become the most frequently used technology to be executed within VLSI chips. The phrase MOS is a reference to the MOSFET's physical structure which includes an electrode with a metal gate that is located on the top of an oxide insulator of semiconductor material.

A material like Aluminum is used only once however the material is now polysilicon. The designing of other metal gates can be done using a comeback through the arrival of high-κ dielectric materials within the process of the CMOS process.

CCD Vs. CMOS

The image sensors like the charge-coupled device (CCD) & complementary metal-oxide-semiconductor (CMOS) are two different kinds of technologies. These are used to capture the image digitally. Every image sensor has its advantages, disadvantages & applications. The main difference between CCD & CMOS is the way of capturing the frame. A charge-coupled device like CCD uses a global shutter whereas the CMOS uses a rolling shutter. These two image sensors change the charge from light to electrical & process it into electronic signals.

The manufacturing process used in CCDs is special to form the capacity to move charge across the IC without alteration. So, this manufacturing process can lead to extremely high-quality sensors about light sensitivity & fidelity. In contrast, CMOS chips use fixed manufacturing procedures to design the chip and a similar process can also be used in making the microprocessors. Because of the differences in manufacturing, there are some clear dissimilarity among the sensors like CCD 7 CMOS.

CCD sensors will capture the images with less noise and huge quality whereas the CMOS sensors are usually more liable to noise. Usually, CMOS uses less power whereas the CCD uses lots of power like more than 100 times to CMOS sensor. The fabrication of CMOS chips can be done on any typical Si production line because they tend to be very cheap as compared with CCDs. CCD sensors are more mature because they are mass-produced for a long period.

Both the CMOS & CCD imagers depend on the effect of photoelectric to make the electrical signal from the light. CCDs are used in cameras to target high-quality images through lots of pixels & outstanding light sensitivity. Usually, CMOS sensors have less resolution, quality & sensitivity. In some applications, CMOS sensors are recently improving to the point wherever they attain near equality with CCD devices. Generally, CMOS cameras are not expensive & they have a high life of the battery.

Latch-up in CMOS

A latch-up can be defined as when the short circuit occurs between the two terminals like power and ground so that high current can be generated & IC can be damaged. In CMOS, latch-up is the occurrence of low impedance trail among the power rail & ground rail because of the communication between the two transistors like parasitic PNP & NPN transistors.

In the CMOS circuit, the two transistors like PNP & NPN is connected to two supply rails like VDD & GND. The protection of these transistors can be done through resistors. In a latch-up transmission, the current will flow from VDD to GND straight through the two transistors so that a short circuit can occur, thus extreme current will flow from VDD to the ground terminal. There are different methods for latch-up prevention:

- In latch-up prevention, high resistance can be placed in the trail to stop the flow of current throughout supply & to make $\beta1*\beta2$ below 1 by using the following methods.
- The structure of parasitic SCR will be beaked in the surrounding of transistors like PMOS & NMOS through an insulating oxide layer. The technology for latch-up protection will turn off the device once latch-up is noticed.
- The testing services of latch-up can be done by many vendors in the market. This test can be done by a sequence of attempts to activate the structure of SCR in the CMOS IC whereas the related pins are checked when overcurrent flows through it.
- It is advised to obtain the first samples from the experimental lot & send them to a testing lab of Latch-up. This lab will apply the utmost achievable power supply & then provide the current supply to the inputs & outputs of the chip whenever a Latch-up occurs through monitoring the current supply.

Advantages

The main benefits of CMOS over TTL are good noise margin as well as less power consumption. This is due to no straight conducting lane from VDD to GND, fall times based on the conditions of input, then the transmission of the digital signal will become easy & low cost through CMOS chips.

CMOS is used to explain the amount of memory on the motherboard of the computer that will store in the settings of BIOS. These settings mainly include the date, time, and settings of hardware. TTL is a digital logic circuit where bipolar transistors work on DC pulses. Several transistor logic gates are normally made-up of a single IC.

The outputs if CMOS drive actively in both ways:

- It uses a single power supply like + VDD.
- These gates are very simple.
- Input impedance is high.
- CMOS logic uses less power whenever it is held in a set state.
- Power dissipation is negligible.
- Fan out is high.
- TTL compatibility.

- Stability of temperature.
- Noise immunity is good.
- Compact.
- Designing is very well.
- Robust mechanically.
- Logic swing is large (VDD).

Disadvantages

The disadvantages of CMOS include the following:

- The cost will be increased once the processing steps increases, however, it can be resolved.
- The packing density of CMOS is low as compared with NMOS.
- MOS chips should be secured from getting static charges by placing the leads shorted otherwise; the static charges obtained within leads will damage the chip. This problem can be solved by including protective circuits otherwise devices.
- Another drawback of the CMOS inverter is that it utilizes two transistors as opposed to one NMOS to build an inverter, which means that the CMOS uses more space over the chip as compared with the NMOS. These drawbacks are small due to the progress within the CMOS technology.

CMOS Applications

Complementary MOS processes were widely implemented and have fundamentally replaced NMOS and bipolar processes for nearly all digital logic applications. CMOS technology has been used for the following digital IC designs:

- Computer memories, CPUs.
- Microprocessor designs.
- Flash memory chip designing.
- Used to design application-specific integrated circuits (ASICs).

3

Types of Logic Gates

Some of the different categories of logic gates are NOT, AND, OR, NAND, NOR, XOR and XNOR gates. The aim of this chapter is to explore different types of logic gates which are closely related to its progress. These topics are crucial for a complete understanding of logic gates.

Logic gates are the basic components in digital electronics. These gates are used to create digital circuits right from simple to complex logic circuit and even complex integrated circuits. Complex microprocessor or microcontroller ICs are constructed using many logic gates. Logic gates are the fundamental building blocks of all digital systems. It has one or more inputs and one output with some logical relationship between them. Logic gate accepts binary signals i.e. True or False, ON or OFF, 1 or 0 and have an ability to make decisions. The state of the output is decided by the input states. All logic gates implements some Boolean function which correlates output with input through some logical operation. Logic gates are mainly designed with the electronic switches using diodes and transistors.

There are three basic gates – AND, OR and NOT. NAND and NOR gates are derived gates are also known as universal logic gates. An XOR gate is inequality detector gate and can be used in comparator, adders, parity generators etc. The symbols of logic gates are of two types – Distinctive shapes & rectangular shapes.

Digital systems are constructed using logic gates. The logic gate is the most basic building block of any digital system capable of making decision including computers. Each one of the basic logic gates is a piece of hardware or an electronic circuit that can be used to implement some basic logic expression. It is an electronic circuit with one or many inputs and only one output.

With the help of logic gates it is possible to implement the most elementary logic expressions, also known as Boolean expressions. The three basic logic gates are the OR gate, the AND gate and the NOT gate. Most logic gates have two inputs and one output. At any given instance, every input or output terminal is in one of the two logic conditions:

- True or false this describes Logic value.
- 1 or 0 provides binary value.
- +5V (+3.3V for new devices) or 0V provides voltage levels.
- High or Low describes the voltage levels.
- ON or OFF describes the switch position.

The common use of logic gate is to act as switches (although they have no moving parts). Gate opens to pass on logic or close to shut to keep it off. This is why they are known as gates. Gates are classified as primary gates (NOT, AND, OR) and Secondary or derived gates (NAND, NOR – Universal gates and XOR, XNOR - combinational gates).

NOT Gate

The NOT gate is a single input single output gate. This gate is also known as Inverter because it performs the inversion of the applied binary signal, i.e., it converts 0 into 1 or 1 into 0. In other words, the gate which has a high input signal only when their input signal is low such type of gate is known as the not gate. The logic symbol for the gate is shown below:

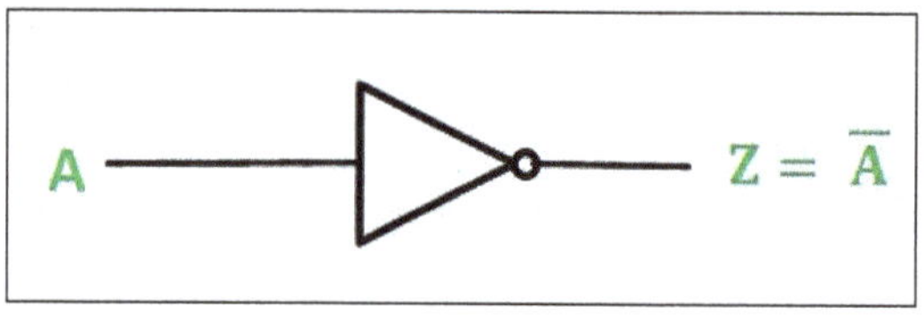

The symbol of the not gate is a triangle with a bubble on its end. This bubble is known as the inversion bubble. It gives the complement of the input signal. It is also known as decision-making devices because it has only one input. The switching circuit diagram of the NOT gate is shown below:

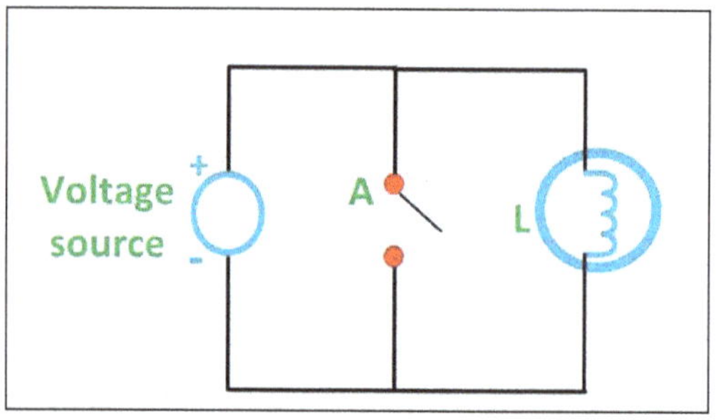

The switching circuit illustrates that the lamp will glow when switch A is open and will go off when the switch A is closed. The truth table of the gate is given below:

A	Z
0	1
1	0

AND Gate

The Logic AND Gate is a type of digital logic circuit whose output goes HIGH to a logic level 1 only when all of its inputs are HIGH. The output state of a digital logic AND gate only returns "LOW" again when ANY of its inputs are at a logic level "0". In other words for a logic AND gate, any LOW input will give a LOW output.

The logic or Boolean expression given for a digital logic AND gate is that for Logical Multiplication which is denoted by a single dot or full stop symbol, (.) giving us the Boolean expression of: A.B = Q. Then we can define the operation of a digital 2-input logic AND gate as being: "If both A and B are true, then Q is true."

2-input Transistor AND Gate

A simple 2-input logic AND gate can be constructed using RTL Resistor-transistor switches connected together as shown below with the inputs connected directly to the transistor bases. Both transistors must be saturated "ON" for an output at Q. Logic AND Gates are available using digital circuits to produce the desired logical function and is given a symbol whose shape represents the logical operation of the AND gate.

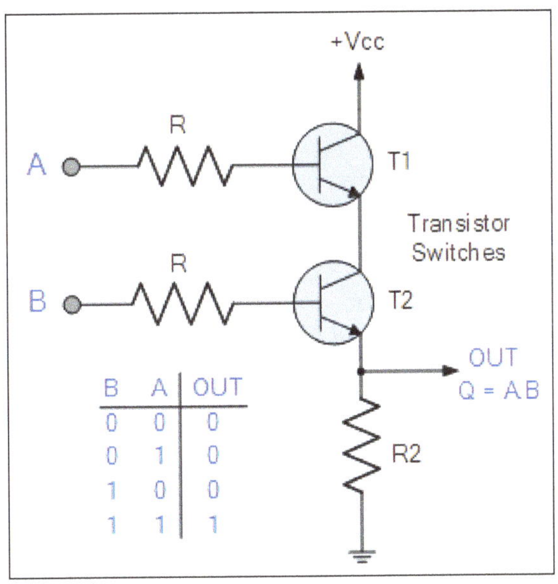

Digital Logic "AND" Gate Types

The 2-input Logic AND Gate

Symbol	Truth Table		
	B	A	Q
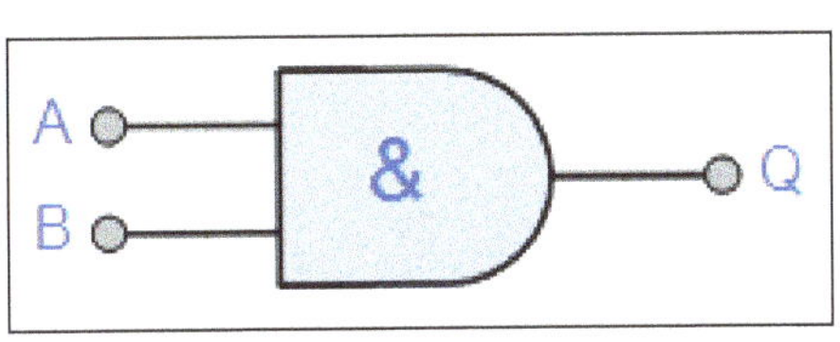 2-input AND Gate.	0	0	0
	0	1	0
	1	0	0
	1	1	1
Boolean Expression Q = A.B	Read as A AND B gives Q		

The 3-input Logic AND Gate

Symbol	Truth Table			
	C	B	A	Q
	0	0	0	0
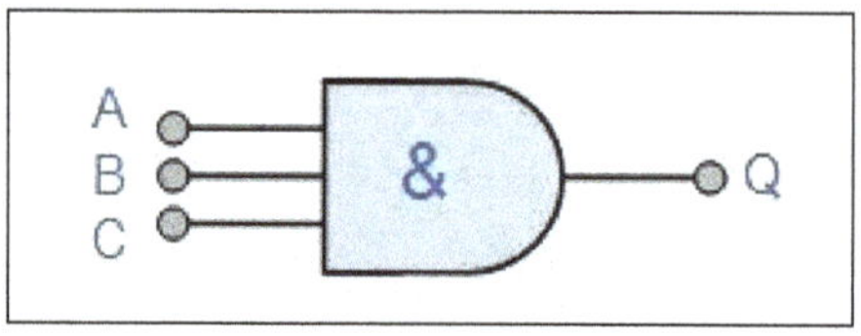	0	0	1	0
	0	1	0	0
	0	1	1	0
	1	0	0	0
3-input AND Gate.	1	0	1	0
	1	1	0	0
	1	1	1	1
Boolean Expression Q = A.B.C	Read as A AND B AND C gives Q			

Because the Boolean expression for the logic AND function is defined as (.), which is a binary operation, AND gates can be cascaded together to form any number of individual inputs. However, commercial available AND gate IC's are only available in standard 2, 3, or 4-input packages. If additional inputs are required, then standard AND gates will need to be cascaded together to obtain the required input value, for example.

Multi-input AND Gate

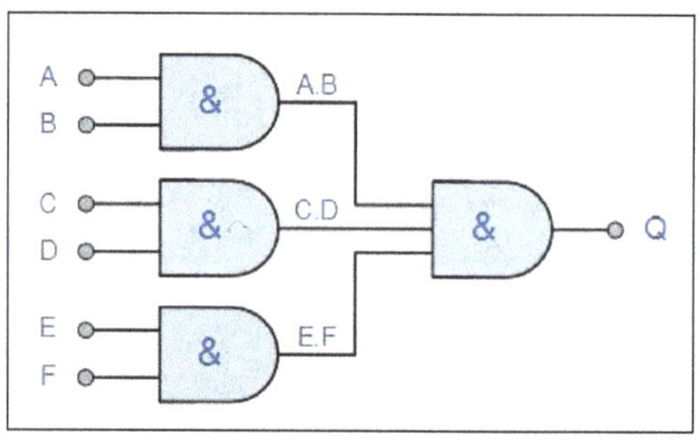

6-input "AND" Function.

The Boolean Expression for this 6-input AND gate will therefore be:

$$Q = (A.B).(C.D).(E.F)$$

In other words: A AND B AND C AND D AND E AND F gives Q.

If the number of inputs required is an odd number of inputs any "unused" inputs can be held HIGH by connecting them directly to the power supply using suitable "Pull-up" resistors.

Commonly available digital logic AND gate IC's include:

- TTL Logic AND Gate:
  - 74LS08 Quad 2-input.
  - 74LS11 Triple 3-input.

- 74LS21 Dual 4-input.

- CMOS Logic AND Gate:
  - CD4081 Quad 2-input.
  - CD4073 Triple 3-input.
  - CD4082 Dual 4-input.

7408 Quad 2-input AND Gate

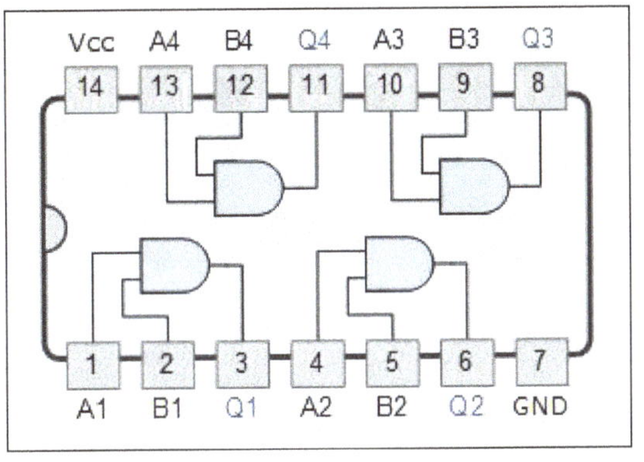

OR Gate

The Logic OR Gate is a type of digital logic circuit whose output goes HIGH to a logic level 1 only when one or more of its inputs are HIGH. The output, Q of a "Logic OR Gate" only returns "LOW" again when ALL of its inputs are at a logic level "0". In other words for a logic OR gate, any "HIGH" input will give a "HIGH", logic level "1" output.

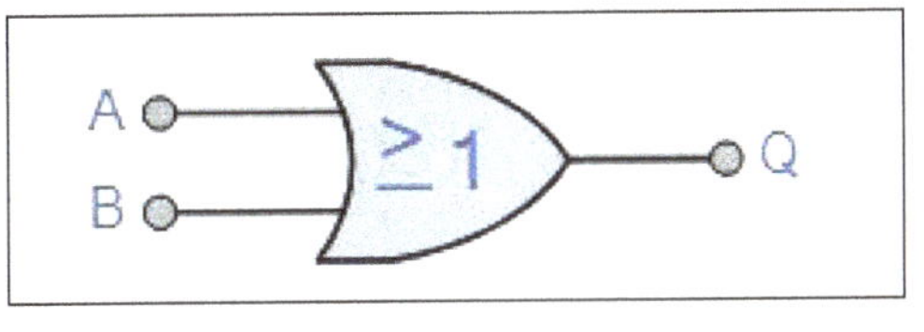

The logic or Boolean expression given for a digital logic OR gate is that for Logical Addition which is denoted by a plus sign, (+) giving us the Boolean expression of: A+B = Q. Thus a logic OR gate can be correctly described as an "Inclusive OR gate" because the output is true when both of its inputs are true (HIGH). Then we can define the operation of a 2-input logic OR gate as being: "If either A or B is true, then Q is true."

2-input Transistor OR Gate

A simple 2-input inclusive OR gate can be constructed using RTL Resistor-transistor switches connected together as shown below with the inputs connected directly to the transistor bases. Either transistor must be saturated "ON" for an output at Q. Logic OR Gates are available using digital circuits to produce the desired logical function and is given a symbol whose shape represents the logical operation of the OR gate.

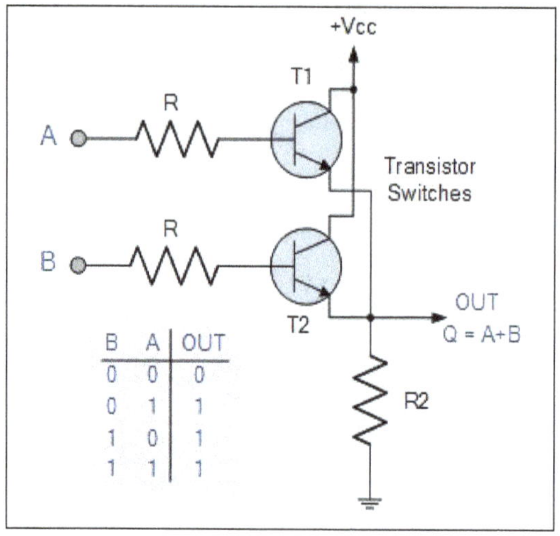

Digital Logic "OR" Gate Types

The 2-input Logic OR Gate

<table>
<tr><td>Symbol</td><td colspan="3">Truth Table</td></tr>
<tr><td rowspan="5">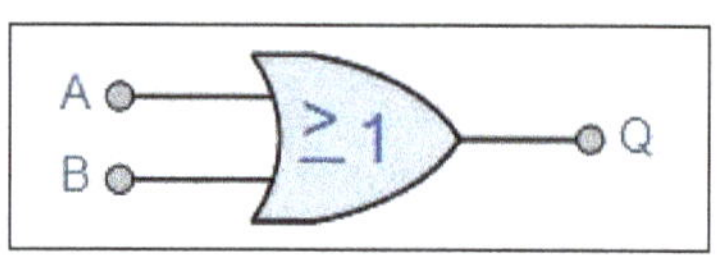2-input OR Gate.</td><td>B</td><td>A</td><td>Q</td></tr>
<tr><td>0</td><td>0</td><td>0</td></tr>
<tr><td>0</td><td>1</td><td>1</td></tr>
<tr><td>1</td><td>0</td><td>1</td></tr>
<tr><td>1</td><td>1</td><td>1</td></tr>
<tr><td>Boolean Expression Q = A+B.</td><td colspan="3">Read as A OR B gives Q.</td></tr>
</table>

The 3-input Logic OR Gate

<table>
<tr><td>Symbol</td><td colspan="4">Truth Table</td></tr>
<tr><td rowspan="9">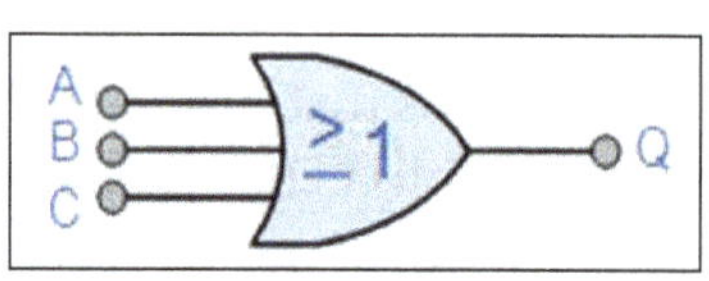3-input OR Gate.</td><td>C</td><td>B</td><td>A</td><td>Q</td></tr>
<tr><td>0</td><td>0</td><td>0</td><td>0</td></tr>
<tr><td>0</td><td>0</td><td>1</td><td>1</td></tr>
<tr><td>0</td><td>1</td><td>0</td><td>1</td></tr>
<tr><td>0</td><td>1</td><td>1</td><td>1</td></tr>
<tr><td>1</td><td>0</td><td>0</td><td>1</td></tr>
<tr><td>1</td><td>0</td><td>1</td><td>1</td></tr>
<tr><td>1</td><td>1</td><td>0</td><td>1</td></tr>
<tr><td>1</td><td>1</td><td>1</td><td>1</td></tr>
<tr><td>Boolean Expression Q = A+B+C.</td><td colspan="4">Read as A OR B OR C gives Q.</td></tr>
</table>

Like the AND gate, the OR function can have any number of individual inputs. However, commercial available OR gates are available in 2, 3, or 4 inputs types. Additional inputs will require gates to be cascaded together for example.

Multi-input OR Gate

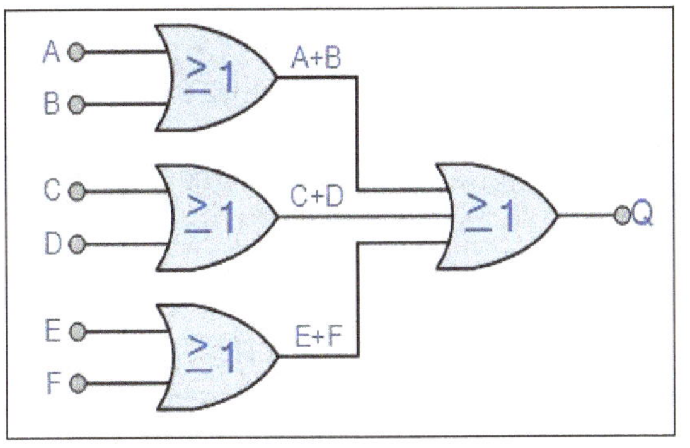

6-input "OR" Function.

The Boolean Expression for this 6-input OR gate will therefore be:

$$Q = (A + B) + (C + D) + (E + F)$$

In other words: A OR B OR C OR D OR E OR F gives Q.

If the number of inputs required is an odd number of inputs any "unused" inputs can be held LOW by connecting them directly to ground using suitable "Pull-down" resistors.

Commonly available digital logic OR gate IC's include:

- TTL Logic OR Gates:
  - 74LS32 Quad 2-input.
- CMOS Logic OR Gates:
  - CD4071 Quad 2-input.
  - CD4075 Triple 3-input.
  - CD4072 Dual 4-input.

7432 Quad 2-input Logic OR Gate

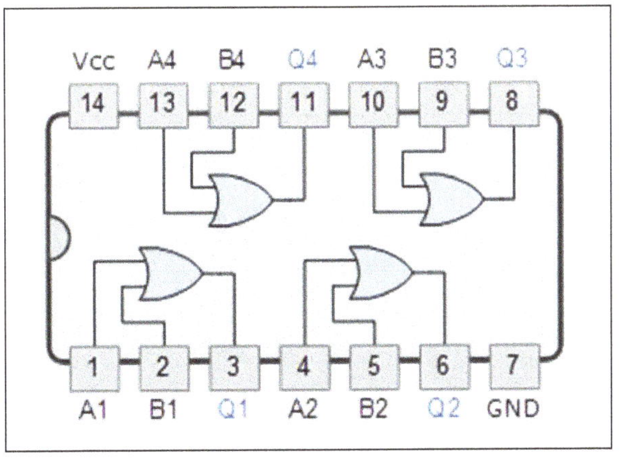

NAND Gate

The NAND gate is a special type of logic gate in the digital logic circuit. The NAND gate is the universal gate. It means all the basic gates such as AND, OR, and NOT gate can be constructed using a NAND gate. The NAND gate is the combination of the NOT-AND gate. The output state of the NAND gate will be low only when all the inputs are high. Simply, this gate returns the complement result of the AND gate.

The logic or Boolean expression for the NAND gate is the complement of logical multiplication of inputs denoted by a full stop or a single dot as (A.B)'=Y. The value of Y will be true when any one of the input is set to 0.

Types of Digital Logic NAND Gate

The NAND gate is also classified into three types based on the input it takes. These are the following types of NAND gate:

The 2-input NAND Gate

This is the simple formation of the NAND gate. In this type of NAND gate, there are only two input values and an output value. There are 2^2=4 possible combinations of inputs. The truth table and logic design are given below:

Logic Design:

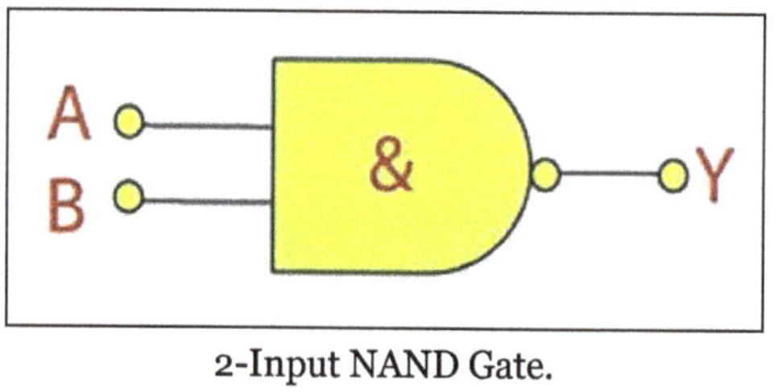

2-Input NAND Gate.

Truth Table:

Input		Output
A	B	Y
0	0	1
0	1	1
1	0	1
1	1	0

The 3-input NAND Gate

Unlike the 2-input NAND gate, the 3-input NAND gate has three inputs. The Boolean expression of the logic NAND gate is defined as the binary operation dot(.). The NAND gate can be cascaded

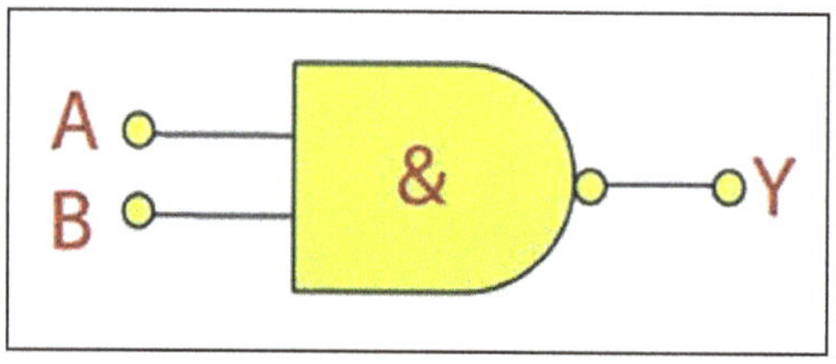

together to form any number of individual inputs. There are 23=8 possible combinations of inputs. The truth table and logic design are given below:

Logic Design:

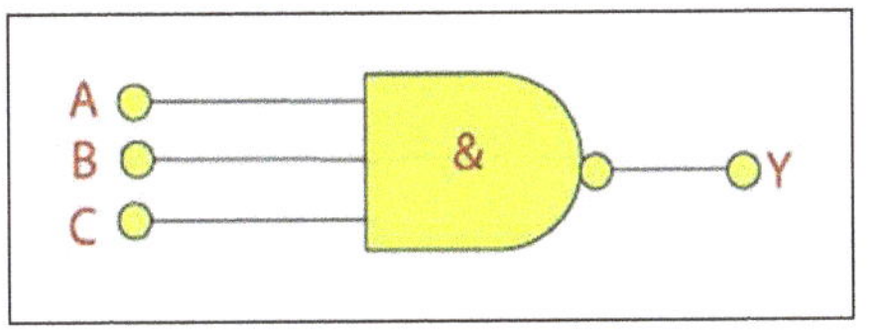

3-Input NAND Gate.

Truth Table:

Input			Output
A	B	C	Y
0	0	0	1
0	0	1	1
0	1	0	1
0	1	1	1
1	0	0	1
1	0	1	1
1	1	0	1
1	1	1	0

The Multi-input NAND Gate

Just like AND, NOT, and OR gate, we can also form n-input NAND gate. If the number of inputs required is odd, any "unused" input can be held high by directly connecting it to the power supply using high "suitable" pull-up resistors. There is the following expression of the 4-input NAND gate:

$$Y = ((A.B).(C.D))'$$

In simple words, it is expressed as: Y = A NAND B NAND C NAND D.

Logic Design:

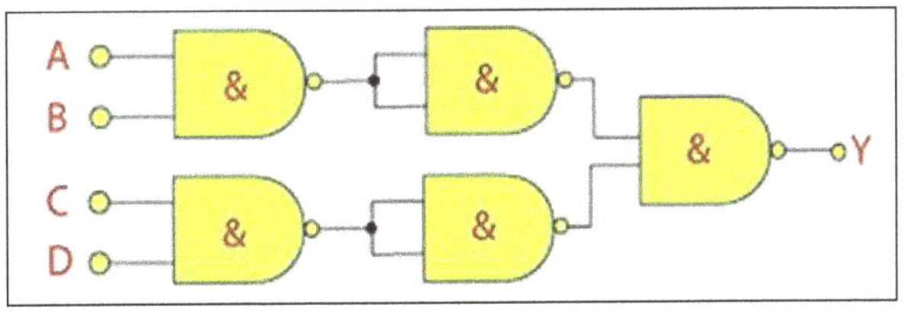

Truth Table:

Input						Output
A	B	C	D	E	F	Y
0	0	0	0	0	0	1

0	0	0	0	0	1	1
0	0	0	0	1	0	1
0	0	0	0	1	1	1
. . .	. . .	. . .	. . .	. . .	. . .	. . .
1	1	1	1	0	0	1
1	1	1	1	0	1	1
1	1	1	1	1	0	1
1	1	1	1	1	1	0

AND Gate using NAND Gate

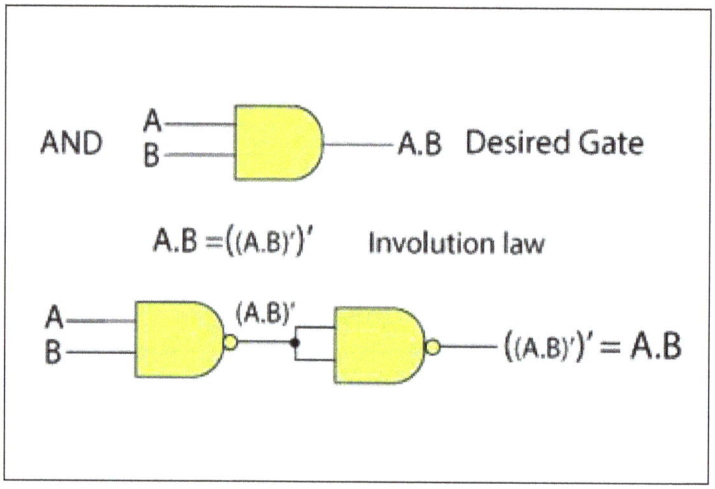

The Derivation of the AND Gate.

OR Gate using NAND Gate

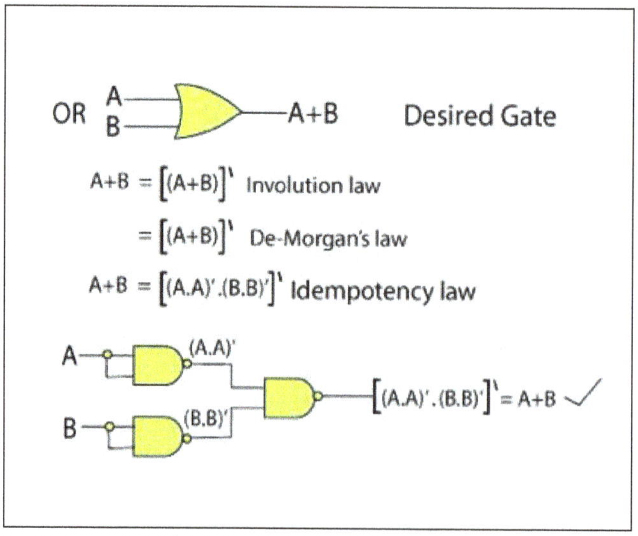

The Derivation of the OR Gate.

NOT Gate using NAND Gate

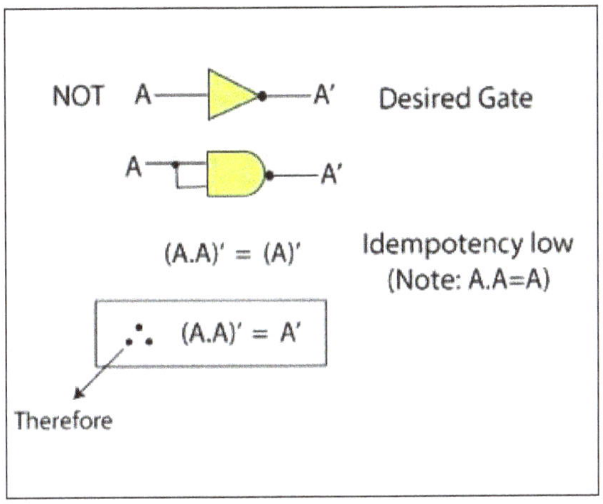

The Derivation of the NOT Gate.

NOR Gate

A NOR gate ("not OR gate") is a logic gate that produces a high output (1) only if all its inputs are false, and low output (0) otherwise. Hence the NOR gate is the inverse of an OR gate, and its circuit is produced by connecting an OR gate to a NOT gate. Just like an OR gate, a NOR gate may have any number of input probes but only one output probe.

A NOT gate followed by an OR gate makes a NOR gate. The basic logic construction of the NOR gate is shown below:

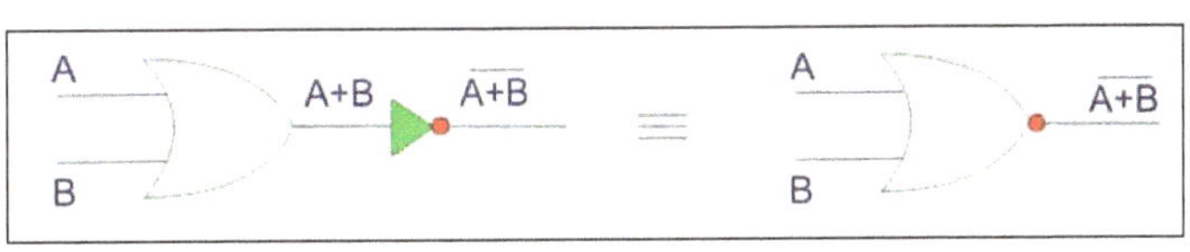

The symbol of the OR gate is similar to the OR gate but one bubble is drawn at the output point of the OR gate in the case of the OR gate.

NOR Gate Truth Table

NOR gate means "not an OR gate" hence the output of this logic gate is just the reverse of that of an OR gate. We know that the output of an OR gate is 0 only when all inputs of the OR gate are 0. Hence in the case of the NOR gate, the output is 1 only when all inputs are 0. In all other cases, that is for all other combinations of inputs the output is 0.

Truth tables list the output of a particular digital logic circuit for all the possible combinations of its inputs. The truth table of a 2 input NOR gate can be represented as:

Inputs		Outputs
A	B	$X = \overline{A + B}$

0	0	1
0	1	0
1	0	0
1	1	0

You can see this is just the reverse of the 2 input OR gate truth table, which is as follows:

Inputs		Outputs
A	B	$X = A + B$
0	0	0
0	1	1
1	0	1
1	1	1

Like the OR gate, the NOR gate may also have more than two inputs.

A NOR gate is also referred to as a universal gate. Because all binary operations can be realized by only using NOR gates. As we know that there are only three basic operations AND, OR, and NOT. Also, we know that all complex binary operations can be realized by using these three basic operations. If we can prove that AND, OR, and NOT operations can be realized by using only the NOR gate, then we can easily say that all binary operations can be realized by using only NOR gates. In digital electronics, other logic gates include NOT gates, OR gates, NAND gates, XOR gates, XNOR gates.

Building a NOT Gate using NOR Gates

If inputs of two inputs OR gate are 1, then the output is 1 and when both inputs of two inputs OR gate is 0, the output is 0. As the NOR gate is reverse of the OR gate when both inputs of the NOR gate are 1, the output will be 0, and when both inputs of two inputs NOR gate 0, the output will be 1.

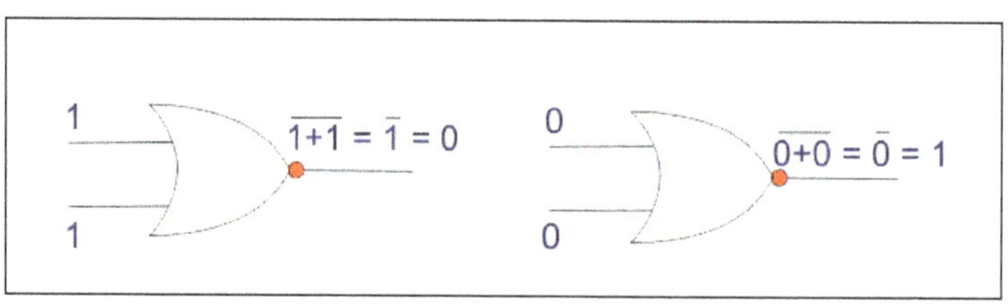

Now if we short circuit both inputs of a NOR gate, then we will get,

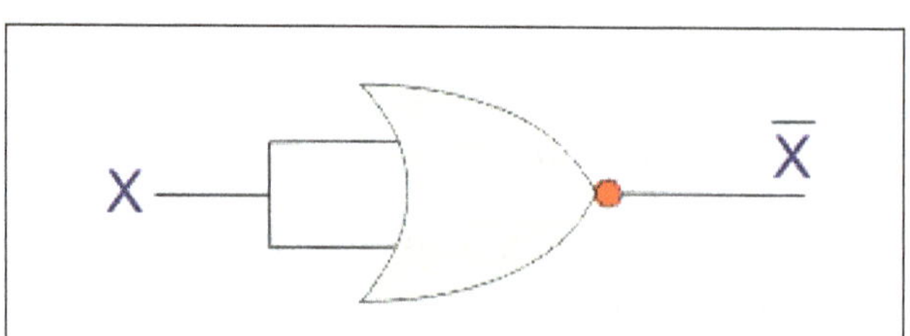

Where X is either 0 or 1. So when X = 1, the output of the above gate is 1. This is nothing but a NOT operation. So we have seen a NOT gate can easily be realized by using the NOR gate only.

Building an OR Gate using NOR Gates

NOR gate is an inverted OR gate, so the inverted NOR gate will be an OR gate. A NOT gate followed by a NOR gate will be equivalent to an OR gate. Here we will use the NOT gate which is realized from another NOR gate. The logic circuit of this OR gate will be as shown below:

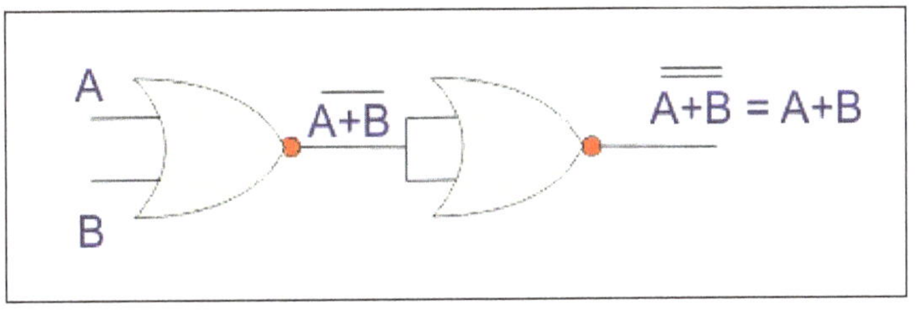

Building an AND Gate from NOR Gate

The NOR operation of two input NOR gate is,

$$\overline{A+B}$$

Now, according to D.Morgan's theorem,

$$\overline{A+B} = \overline{A}.\overline{B}$$

Now, if we put inverted A instead of A and inverted B instead of B in the above equation, we will get,

$$\overline{\overline{A}+\overline{B}} = \overline{\overline{A}}.\overline{\overline{B}} = A.B$$

This is an AND operation.

So from the above equation, it is clear that if inputs of a NOR gate are inverted then the output will be AND equivalent. So, if we connect the NOT gate at both inputs of a NOR gate, we will get an AND gate equivalent. As we are only allowed to use the NOR gate, we will use here the NOT gates which are realized from the NOR gate. The logical equivalent circuit will be as shown below:

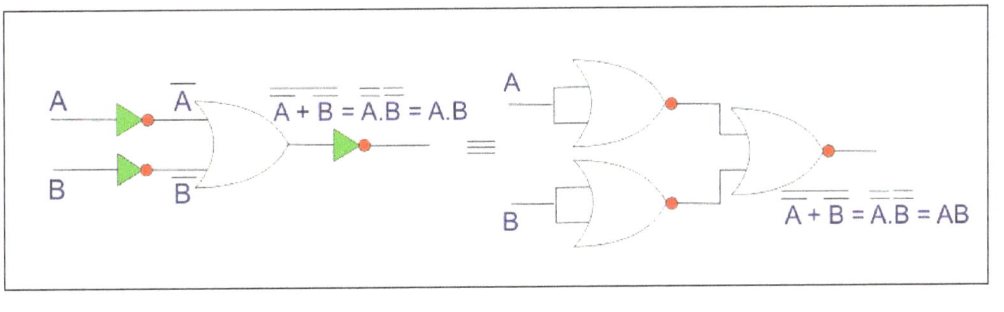

Now we have proved that all three basic logical gates, NOT, OR, and AND gates can be realized by using only NOR gates. So any even much complex binary operation can be realized by using only a NOR gate hence it is quite justified to cal a NOR gate as a universal logic gate.

NOR Gate Circuit Diagram

NOR Gate Transistor Circuit Diagram

A NOR gate can be realized by using two bipolar junction transistors. The basic transistor circuit diagram is shown below. This circuit is made by two parallel-connected transistors.

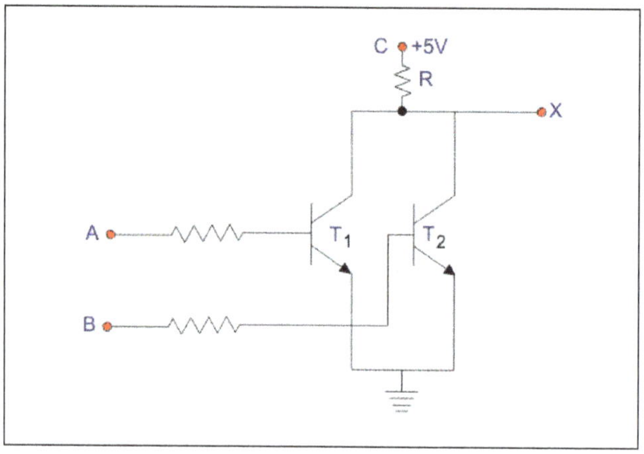

In the circuit when both A and B are given +5 V, base of the both transistors gets quite a high potential to make the transistors 'ON'. As both transistors T_1 and T_2 are in ON condition, supply voltage at terminal C gets the path to the ground through resistor R.

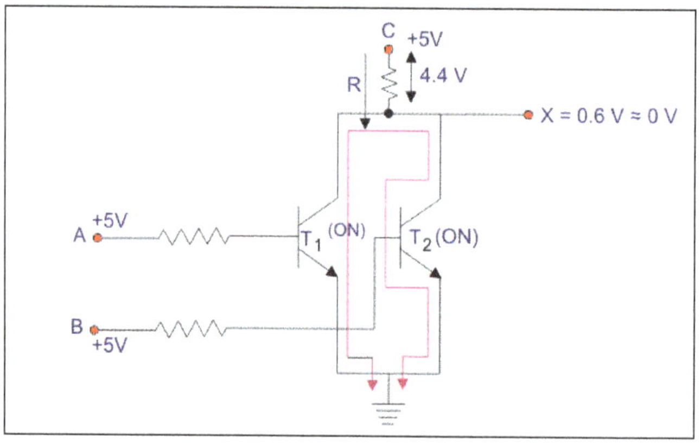

The entire supply voltage ideally drops across resistor R and the output terminal of the circuit will not get any voltage and hence it will be at logical 0 states.

Practically, a transistor cannot be ideally short-circuited in ON condition. There will be a voltage drop between emitter and collector of the transistor even in ON condition. So practically the entire supply voltage will not drop across resistor R instead a small portion of it (0.6 V) is dropped across the transistor (saturated). That is, 5 – 0.6 = 4.4 V. Hence, the voltage appears at output terminal X is not practically zero instead it will be 0.6 V which is considered as logical 0.

Now, if either of inputs A and B are given with +5 V, the only corresponding transistor will be in ON condition. But in this case, also supply voltage will get the path to the ground through R and ON transistor and similarly, the output will be in logical 0 states which is 0.6 V.

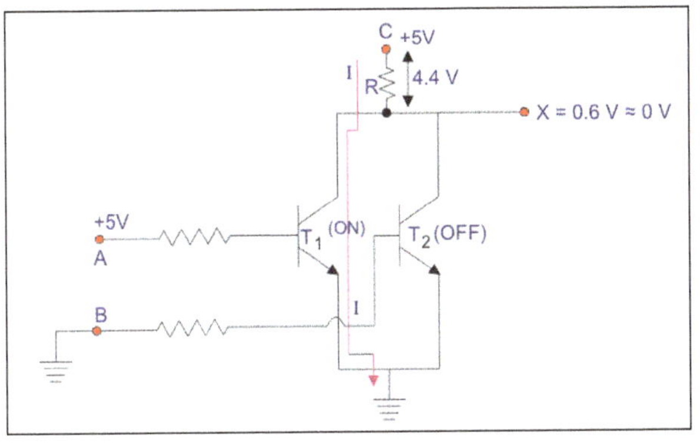

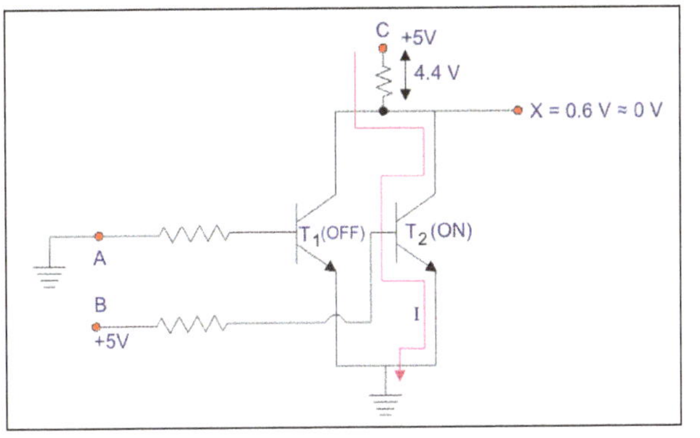

Now if both of the inputs A and B are given with 0V or grounded, both transistors will be in OFF condition as in this case the base of both transistors does not have enough potential to make the transistors ON.

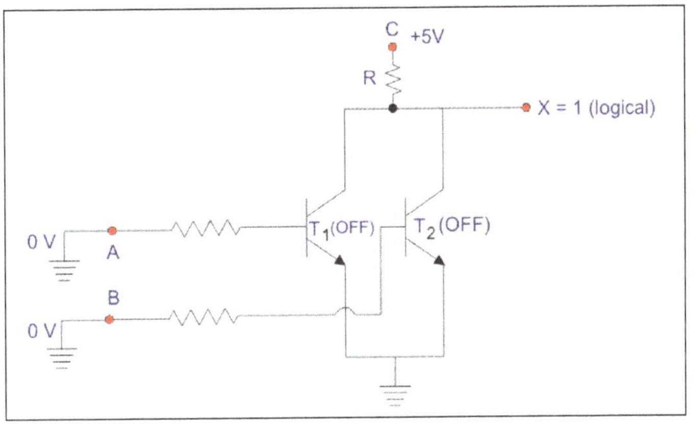

As the supply voltage will not get the nay path to the ground, the supply voltage will appear at output X, hence the output is in logical 1 or high state. Hence, in the circuit, the output is 1, only when both of the inputs are 0 and in all other conditions, the output is 0.

NOR Gate Integrated Circuits

There are many NOR gate Integrated Circuits (ICs) available in the market. A few of the most popular include:

- IC 7402 contains four two inputs NOR gates:

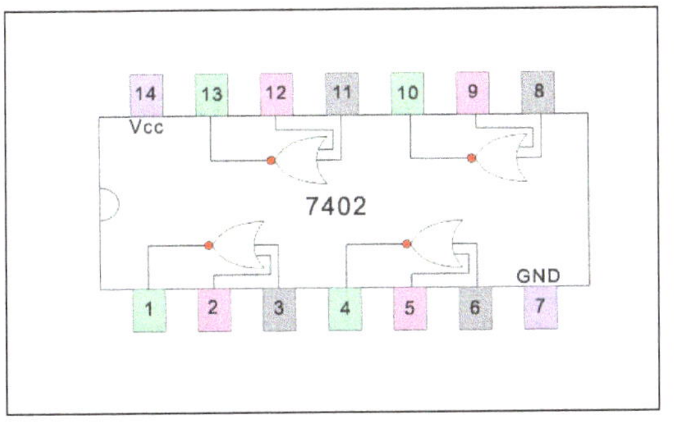

- IC 7427 contains three inputs NOR gates:

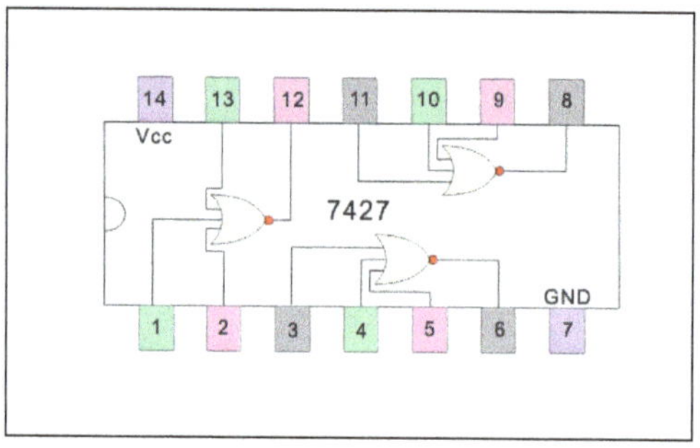

- IC 7425 contains two four inputs NOR gates:

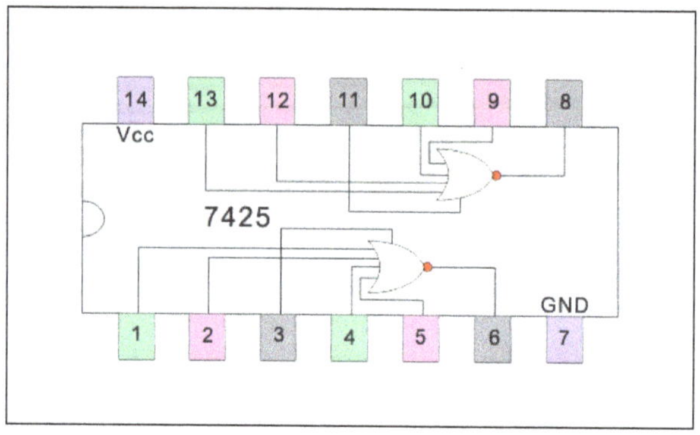

XOR Gate

The Exclusive-OR logic function is a very useful circuit that can be used in many different types of computational circuits. By using the three principal gates, the AND Gate, the OR Gate and the NOT Gate, we can build many other types of logic gate functions, such as a NAND Gate and a NOR Gate or any other type of digital logic function we can imagine.

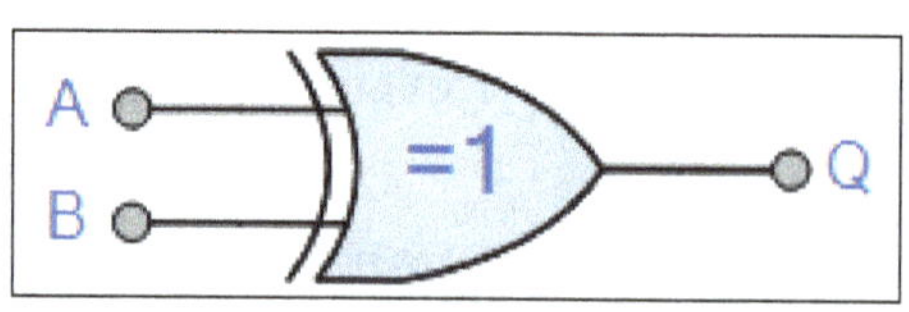

But there are two other types of digital logic gates which although they are not a basic gate in their own right as they are constructed by combining together other logic gates, their output Boolean function is important enough to be considered as complete logic gates. These two "hybrid" logic gates are called the Exclusive-OR (Ex-OR) Gate and its complement the Exclusive-NOR (Ex-NOR) Gate.

For a 2-input OR gate, if A = "1", OR B = "1", OR BOTH A + B = "1" then the output from the digital gate must also be at a logic level "1" and because of this, this type of logic gate is known as an Inclusive-OR function. The logic gate gets its name from the fact that it includes the case of Q = "1" when both A and B = "1".

If however, an logic output “1” is obtained when ONLY A = “1” or when ONLY B = “1” but NOT both together at the same time, giving the binary inputs of “01” or “10”, then the output will be “1”. This type of gate is known as an Exclusive-OR function or more commonly an Ex-Or function for short. This is because its Boolean expression excludes the “OR BOTH” case of Q = “1” when both A and B = “1”. In other words the output of an Exclusive-OR gate ONLY goes “HIGH” when its two input terminals are at “DIFFERENT” logic levels with respect to each other.

An odd number of logic “1’s” on its inputs gives a logic “1” at the output. These two inputs can be at logic level “1” or at logic level “0” giving us the Boolean expression of: $Q = (A \oplus B) = \overline{A}.B + A.\overline{B}$.

The Exclusive-OR Gate function, or Ex-OR for short, is achieved by combining standard logic gates together to form more complex gate functions that are used extensively in building arithmetic logic circuits, computational logic comparators and error detection circuits.

The two-input “Exclusive-OR” gate is basically a modulo two adder, since it gives the sum of two binary numbers and as a result are more complex in design than other basic types of logic gate. The truth table, logic symbol and implementation of a 2-input Exclusive-OR gate is shown below.

The Digital Logic “Exclusive-OR” Gate

2-input Ex-OR Gate

Symbol	Truth Table		
	B	A	Q
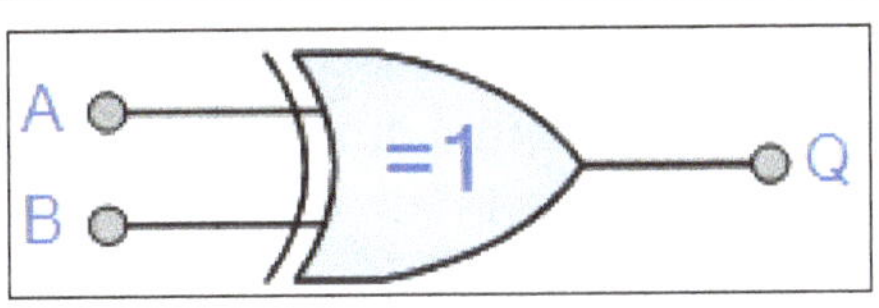	0	0	0
	0	1	1
	1	0	1
2-input Ex-OR Gate.	1	1	0
Boolean Expression Q = A ⊕ B.	A OR B but NOT BOTH gives Q.		

Giving the Boolean expression of: $Q = A\overline{B} + \overline{A}B$.

The truth table above shows that the output of an Exclusive-OR gate ONLY goes “HIGH” when both of its two input terminals are at “DIFFERENT” logic levels with respect to each other. If these two inputs, A and B are both at logic level “1” or both at logic level “0” the output is a “0” making the gate an “odd but not the even gate”. In other words, the output is “1” when there are an odd number of 1’s in the inputs.

This ability of the Exclusive-OR gate to compare two logic levels and produce an output value dependent upon the input condition is very useful in computational logic circuits as it gives us the following Boolean expression of:

$$Q = (A \oplus B) = \overline{A}.B + A.\overline{B}.$$

The logic function implemented by a 2-input Ex-OR is given as either: “A OR B but NOT both” will give an output at Q. In general, an Ex-OR gate will give an output value of logic “1” ONLY when there are an ODD number of 1’s on the inputs to the gate, if the two numbers are equal, the output is “0”.

Then an Ex-OR function with more than two inputs is called an "odd function" or modulo-2-sum (Mod-2-SUM), not an Ex-OR. This description can be expanded to apply to any number of individual inputs as shown below for a 3-input Ex-OR gate.

3-input Ex-OR Gate

<table>
<tr><td>Symbol</td><td colspan="4">Truth Table</td></tr>
<tr><td rowspan="9">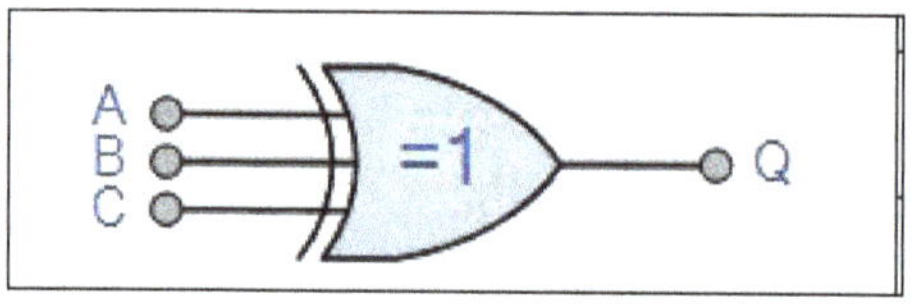
3-input Ex-OR Gate.</td><td>C</td><td>B</td><td>A</td><td>Q</td></tr>
<tr><td>0</td><td>0</td><td>0</td><td>0</td></tr>
<tr><td>0</td><td>0</td><td>1</td><td>1</td></tr>
<tr><td>0</td><td>1</td><td>0</td><td>1</td></tr>
<tr><td>0</td><td>1</td><td>1</td><td>0</td></tr>
<tr><td>1</td><td>0</td><td>0</td><td>1</td></tr>
<tr><td>1</td><td>0</td><td>1</td><td>0</td></tr>
<tr><td>1</td><td>1</td><td>0</td><td>0</td></tr>
<tr><td>1</td><td>1</td><td>1</td><td>1</td></tr>
<tr><td>Boolean Expression Q = A ⊕ B ⊕ C.</td><td colspan="4">"Any ODD Number of Inputs" gives Q.</td></tr>
</table>

Giving the Boolean expression of: $Q = A\overline{BC} + \overline{A}B\overline{C} + \overline{AB}C + ABC.$

The symbol used to denote an Exclusive-OR odd function is slightly different to that for the standard Inclusive-OR Gate. The logic or Boolean expression given for a logic OR gate is that of logical addition which is denoted by a standard plus sign. The symbol used to describe the Boolean expression for an Exclusive-OR function is a plus sign, (+) within a circle (O). This exclusive-OR symbol also represents the mathematical "direct sum of sub-objects" expression, with the resulting symbol for an Exclusive-OR function being given as: (⊕).

The Ex-OR function is not a basic logic gate but a combination of different logic gates connected together. Using the 2-input truth table above, we can expand the Ex-OR function to: $(A+B).(A.B)$ which means that we can realize this new expression using the following individual gates.

Ex-OR Gate Equivalent Circuit

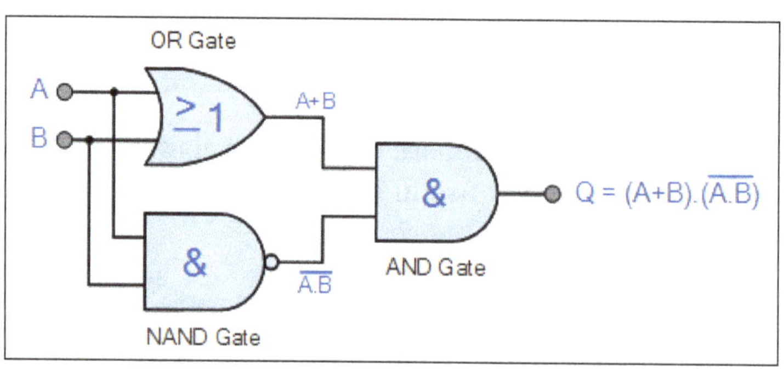

One of the main disadvantages of implementing the Ex-OR function above is that it contains three different types logic gates OR, NAND and finally AND within its design. One easier way of producing the Ex-OR function from a single gate is to use our old favourite the NAND gate.

Ex-OR Function Realisation using NAND Gates

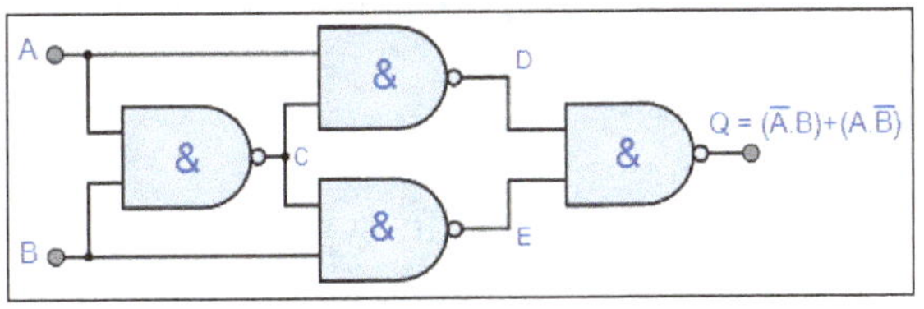

NAND Gate Realisation.

Exclusive-OR Gates are used mainly to build circuits that perform arithmetic operations and calculations especially Adders and Half-Adders as they can provide a "carry-bit" function or as a controlled inverter, where one input passes the binary data and the other input is supplied with a control signal. Commonly available digital logic Exclusive-OR gate IC's include:

- TTL Logic Ex-OR Gates: 74LS86 Quad 2-input.
- CMOS Logic Ex-OR Gates: CD4030 Quad 2-input.

7486 Quad 2-input Exclusive-OR Gate

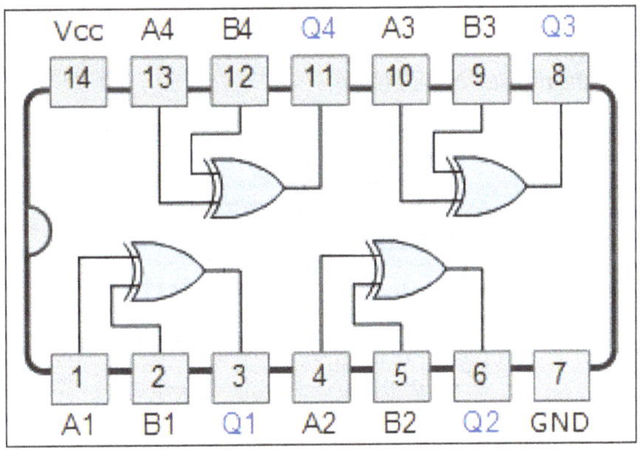

The Exclusive-OR logic function is a very useful circuit that can be used in many different types of computational circuits. Although not a basic logic gate in its own right, its usefulness and versatility has turned it into a standard logical function complete with its own Boolean expression, operator and symbol. The Exclusive-OR Gate is widely available as a standard quad two-input 74LS86 TTL gate or the 4030B CMOS package.

One of its most commonly used applications is as a basic logic comparator which produces a logic "1" output when its two input bits are not equal. Because of this, the exclusive-OR gate has an inequality status being known as an odd function. In order to compare numbers that contain two or more bits, additional exclusive-OR gates are needed with the 74LS85 logic comparator being 4-bits wide.

XNOR Gate

The XNOR gate is the complement of the XOR gate. It is a hybrid gate. Simply, it is the combination of the XOR gate and NOT gate. The output level of the XNOR gate is high only when both of its inputs are the same, either 0 or 1. The symbol of the XNOR gate is the same as XOR, only complement sign is added. Sometimes, the XNOR gate is also called the Equivalence gate.

2-input Ex-NOR Gate

It is a simple form of the hybrid gate XNOR. In this type of XNOR gate, there are only two input values and an output value. There are $2^2=4$ possible combinations of inputs. The output level is high when both inputs are set to high(1). The Boolean expression of 2-input XNOR gate is as follows:

$$Y = (A \oplus B)'$$
$$Y = ((AB)' + AB)$$

Logic Design:

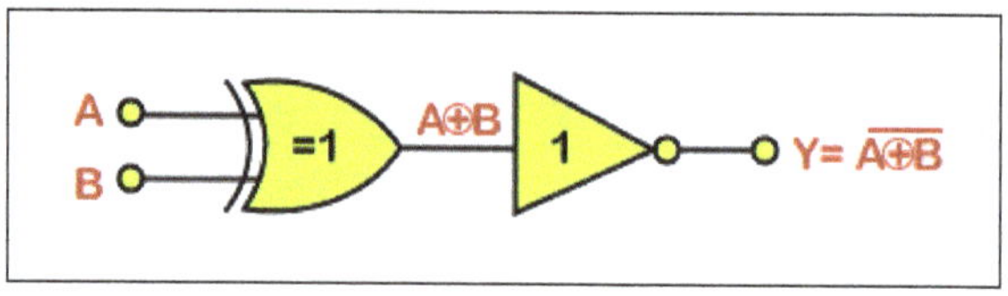

2-input "Ex-OR" Gate plus a "NOT" Gate.

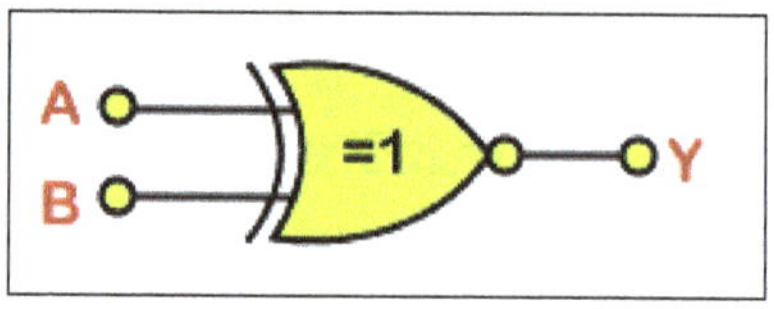

2-input "Ex-NOR" Gate.

Truth Table:

Input		Output
A	B	Y
0	0	1
0	1	0
1	0	0
1	1	1

The 3-Input XNOR Gate

Unlike the 2-input XNOR gate, the 3-input XNOR gate has three inputs. There are $2^3=8$ possible combinations of inputs. The Boolean expression of the logical Ex-OR gate is as follows:

$$Y = (A \oplus B \oplus C)'$$
$$Y = (ABC)' + ABC' + AB'C + A'BC$$

Logic Design:

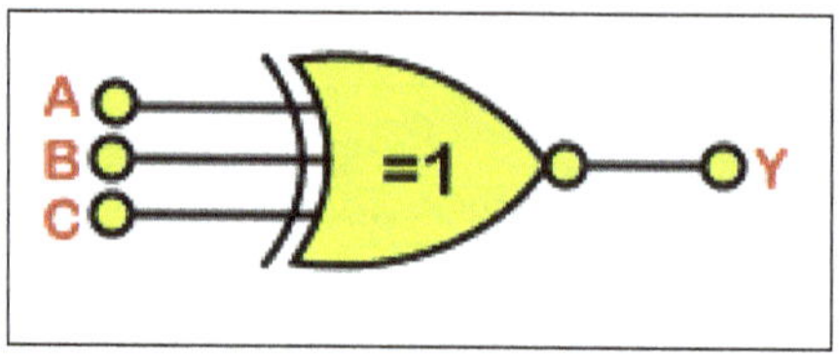

3-input Ex-NOR Gate.

Truth Table:

Input			Output
A	B	C	Y
0	0	0	1
0	0	1	0
0	1	0	0
0	1	1	1
1	0	0	0
1	0	1	1
1	1	0	1
1	1	1	0

Ex-OR Gate Equivalent Circuit

We can form the XNOR or Ex-NOR gate using the gates such as AND, OR, and NOT gate. The main disadvantage of this implementation is that we use different types of gates to form a single XNOR gate. By using the NAND gates only, we can implement the Ex-NOR gate also. This is an easier way of producing Ex-NOR gate functionality.

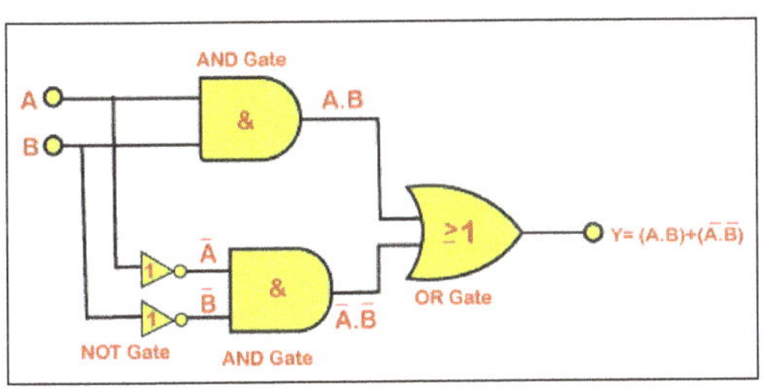

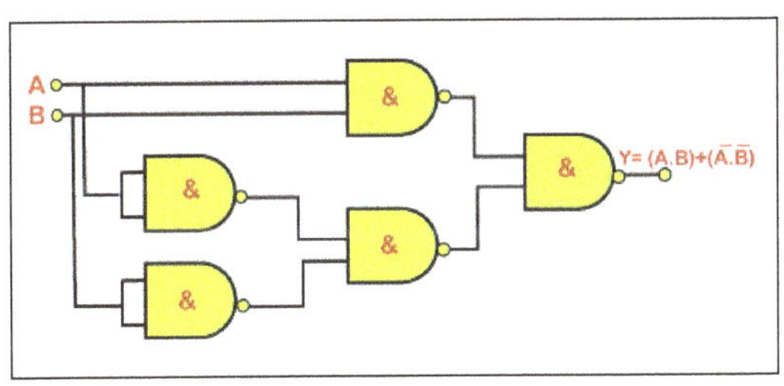

NAND Gate realisation.

Use of Ex-OR Gate

Ex-NOR gates are used mainly in electronic circuits that perform arithmetic operations and data checking such as Adders, Subtractors or Parity Checkers, etc. As the Ex-NOR gate gives an output of logic level "1," whenever its two inputs are equal, it can be used to compare the magnitude of two binary digits or numbers and so Ex-NOR gates are used in Digital Comparator circuits.

Universal Logic Gates

A universal gate is a gate which can implement any Boolean function without need to use any other gate type. The NAND and NOR gates are universal gates. In practice, this is advantageous since NAND and NOR gates are economical and easier to fabricate and are the basic gates used in all IC digital logic families. In fact, an AND gate is typically implemented as a NAND gate followed by an inverter not the other way around. Likewise, an OR gate is typically implemented as a NOR gate followed by an inverter not the other way around.

NAND Gate is a Universal Gate

To prove that any Boolean function can be implemented using only NAND gates, we will show that the AND, OR, and NOT operations can be performed using only these gates.

1. Implementing an Inverter Using only NAND Gate: The figure shows two ways in which a NAND gate can be used as an inverter (NOT gate).

- All NAND input pins connect to the input signal A gives an output A.

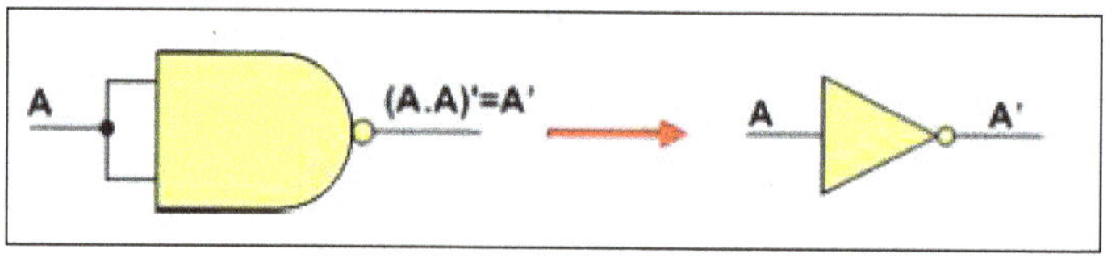

- One NAND input pin is connected to the input signal A while all other input pins are connected to logic 1. The output will be A.

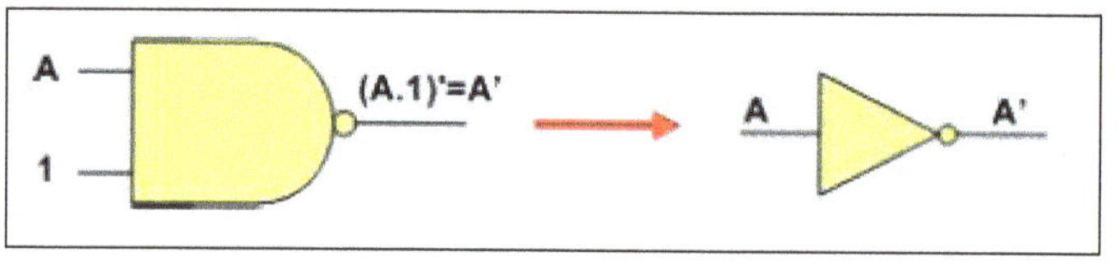

2. Implementing AND Using only NAND Gates: An AND gate can be replaced by NAND gates as shown in the figure (The AND is replaced by a NAND gate with its output complemented by a NAND gate inverter).

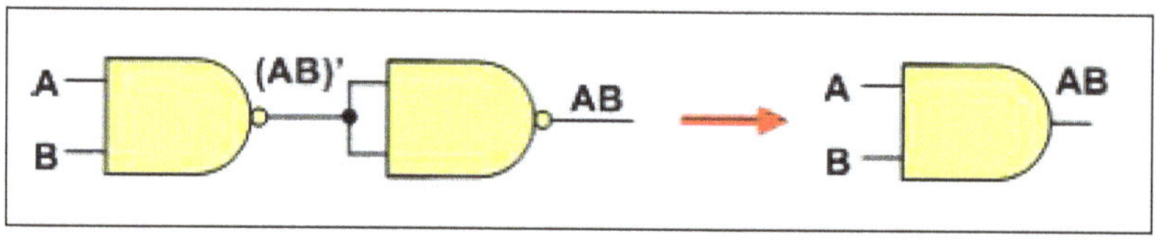

3. Implementing OR Using only NAND Gates: An OR gate can be replaced by NAND gates as shown in the figure (The OR gate is replaced by a NAND gate with all its inputs complemented by NAND gate inverters).

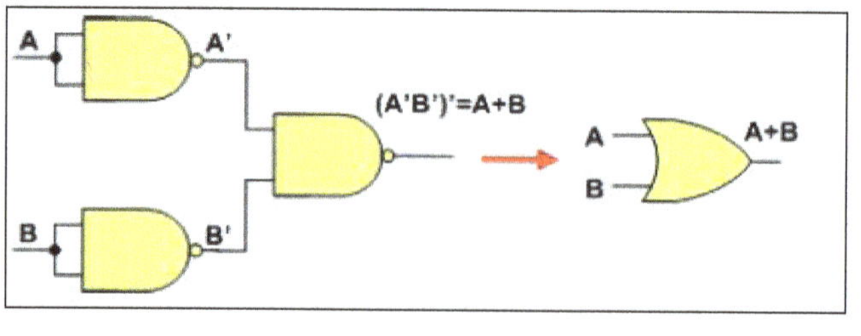

Thus, the NAND gate is a universal gate since it can implement the AND, OR and NOT functions.

Equivalent Gates

The shown figure shows important cases of gate equivalence. Note that bubbles indicate a complement operation (inverter).

1. A NAND gate is equivalent to an inverted-input OR gate.

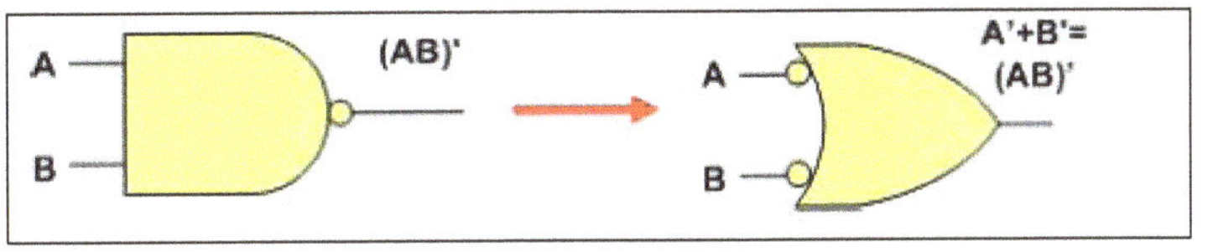

2. An AND gate is equivalent to an inverted-input NOR gate.

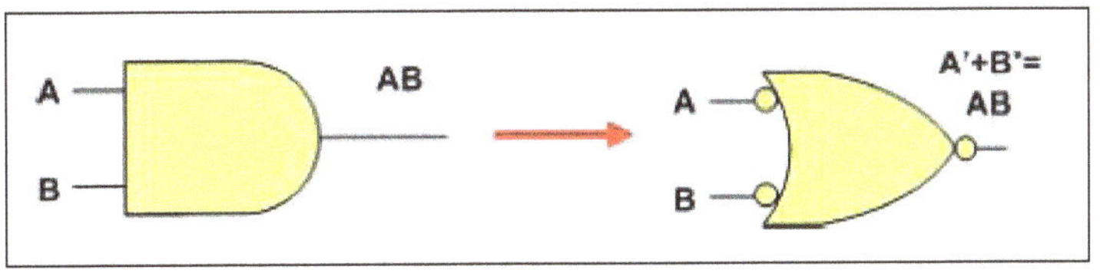

3. A NOR gate is equivalent to an inverted-input AND gate.

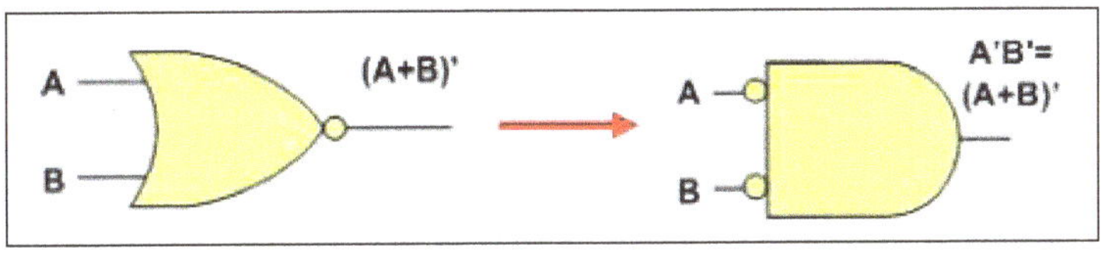

4. An OR gate is equivalent to an inverted-input NAND gate.

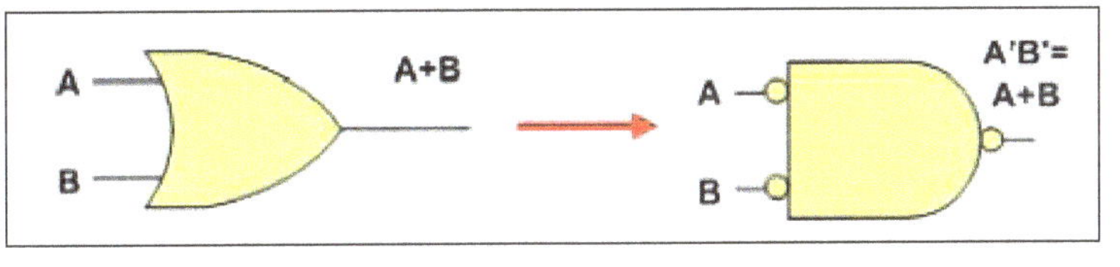

5. Two NOT gates in series are same as a buffer because they cancel each other as A" = A.

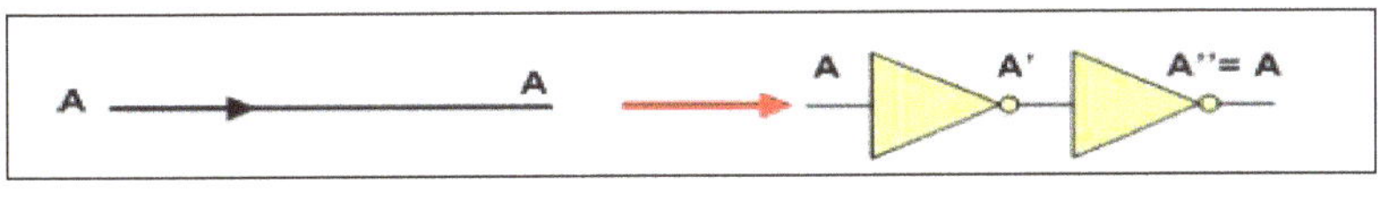

Two-Level Implementations

Boolean functions in either SOP or POS forms can be implemented using 2-Level implementations.

- For SOP forms AND gates will be in the first level and a single OR gate will be in the second level.
- For POS forms OR gates will be in the first level and a single AND gate will be in the second level.

Using inverters to complement input variables is not counted as a level. We will show that SOP forms can be implemented using only NAND gates, while POS forms can be implemented using only NOR gates. This is best explained through examples.

Example: Implement the following SOP function,

$$F = XZ + Y'Z + X'YZ$$

Being an SOP expression, it is implemented in 2-levels as shown in the figure.

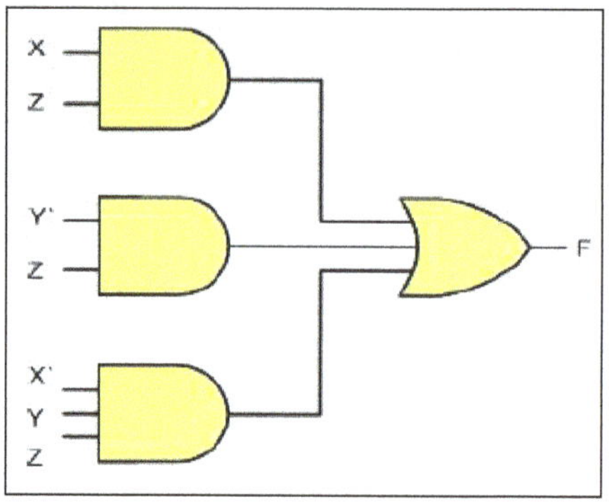

Introducing two successive inverters at the inputs of the OR gate results in the shown equivalent implementation. Since two successive inverters on the same line will not have an overall effect on the logic.

By associating one of the inverters with the output of the first level AND gate and the other with the input of the OR gate, it is clear that this implementation is reducible to 2-level implementation where both levels are NAND gates as shown in Figure.

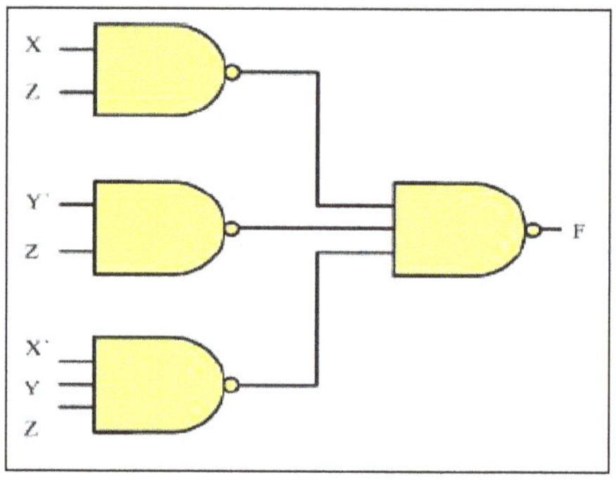

Example: Implement the following POS function,

$$F=(X+Z)(Y'+Z)(X'+Y+Z)$$

Being a POS expression, it is implemented in 2-levels as shown in the figure.

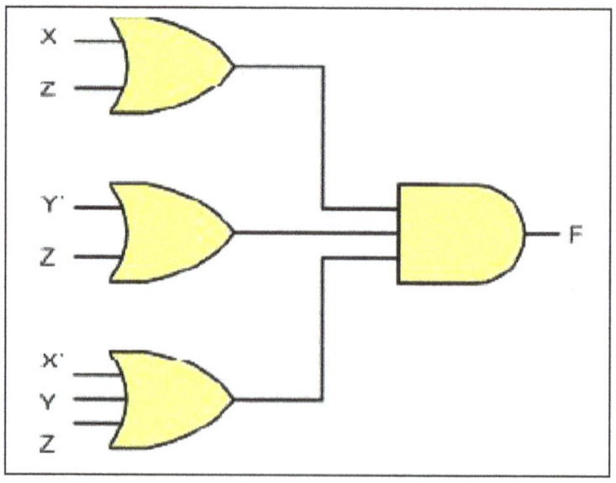

Introducing two successive inverters at the inputs of the AND gate results in the shown equivalent implementation. Since two successive inverters on the same line will not have an overall effect on the logic.

By associating one of the inverters with the output of the first level OR gates and the other with the input of the AND gate, it is clear that this implementation is reducible to 2-level implementation where both levels are NOR gates as shown in Figure.

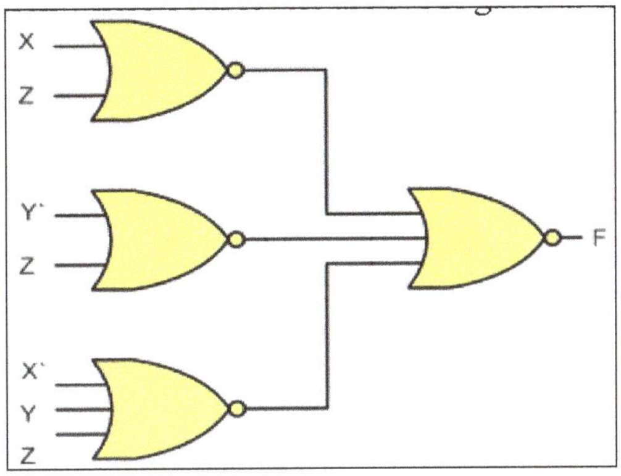

There are some other types of 2-level combinational circuits which are:

- NAND-AND
- AND-NOR
- NOR-OR
- OR-NAND

AND-NOR Functions

Example: Implement the following function,

$$F = \overline{XZ + \overline{Y}Z + \overline{X}YZ} \text{ or}$$

$$F = XZ + \overline{Y}Z + \overline{X}YZ$$

Since F' is in SOP form, it can be implemented by using NAND-NAND circuit. By complementing the output we can get F, or by using NAND-AND circuit as shown in the figure.

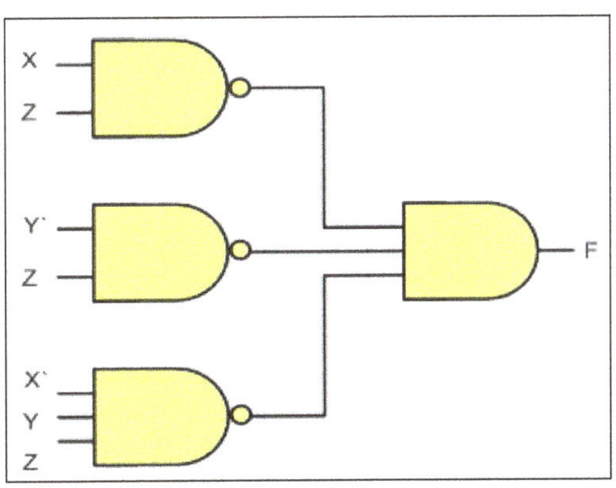

It can also be implemented using AND-NOR circuit as it is equivalent to NANDAND circuit as shown in the figure.

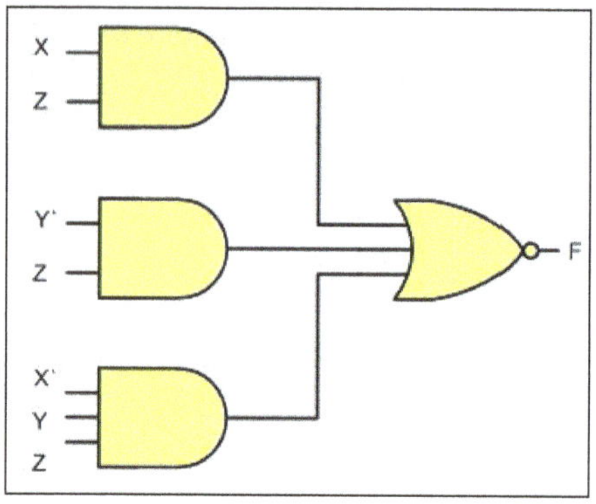

OR-NAND Functions

Example: Implement the following function,

$$F=\overline{(X+Z).(\overline{Y}+Z).(\overline{X}+Y+Z)} \text{ or}$$

$$\overline{F}=(X+Z)(\overline{Y}+Z)(\overline{X}+Y+Z)$$

Since F' is in POS form, it can be implemented by using NOR-NOR circuit. By complementing the output we can get F, or by using NOR-OR circuit as shown in the figure.

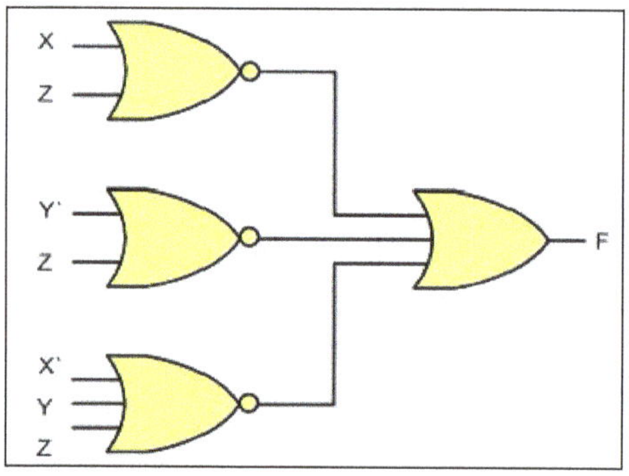

It can also be implemented using OR-NAND circuit as it is equivalent to NOR-OR circuit as shown in the figure.

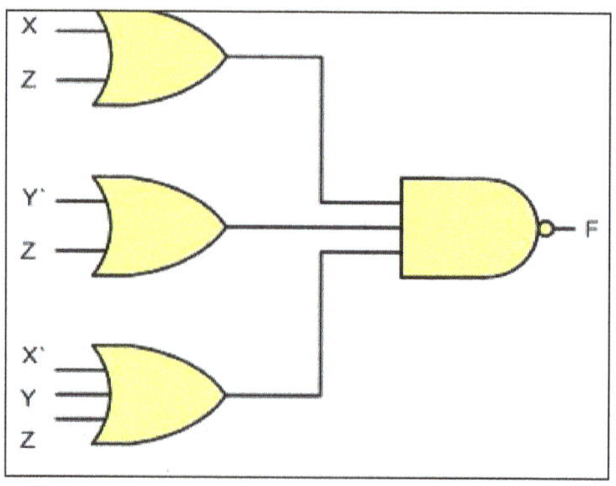

Special Logic Gates

Gates with Open Collector

In case of TTL logic family, some NAND require an external resistor, called the pull-up resistor. Pull up resistor us connected between the output and the DC power supply to get desired logic function. In case of MOS family open drain gates are used. These logic gates have capability of providing an ANDing operation when outputs without using an AND gate. This connection is popularly known as wire-AND system. Figure indicates Wired AND gate using open collector NAND gates. The output of wire-AND system is,

$$Y = AB'. CD'.EF'.$$

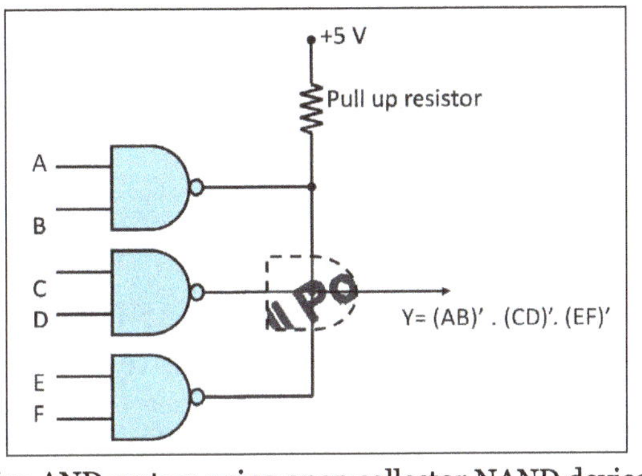

Wire-AND system using open collector NAND devices.

The open collector outputs of several NAND gates can be connected together to form then Wired-AND function. In this example, three open collector NAND gate outputs are wired together as shown in this figure. One single pull up resistor is sufficient for this interconnection. Figure shows similar circuit configuration for Wire-ANDing using open collector NOT gates.

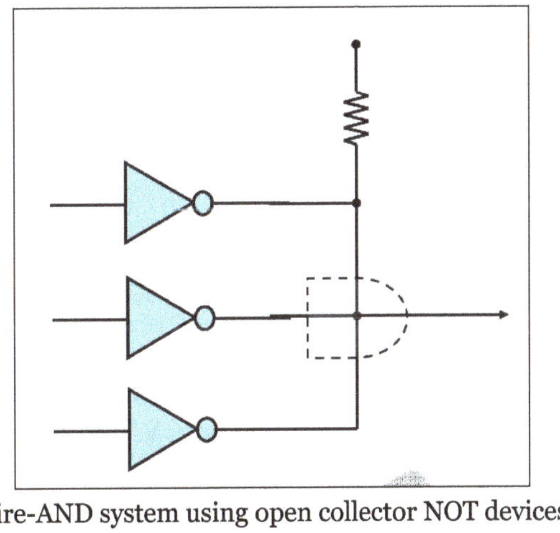

Wire-AND system using open collector NOT devices.

The output Y of wire-ANDing of all the inputs is,

$$Y = (A)'.(B)'.(C)'$$

Tristate Gates

In digital electronics, tri-state (three states) logic allows three possible output states namely the logic '1' state, the logic '0' state and a high-impedance state. This logic gate allows sharing same output line or lines. The high-impedance state is controlled by an external ENABLE input. The ENABLE input decides whether the tri-state gate is active (0 or 1) or in the high-impedance state.

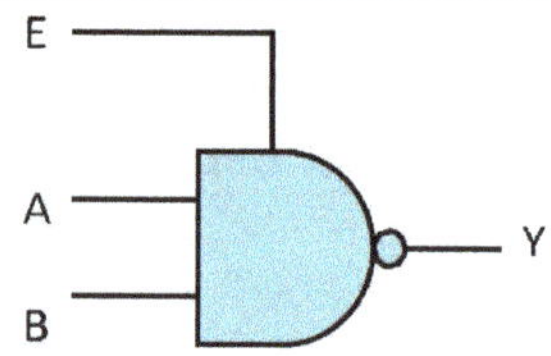

E	A	B	Y
0	0	0	Z
0	0	1	Z
0	1	0	Z
0	1	1	Z
1	0	0	1
1	0	1	1
1	1	0	1
1	1	1	0

Tri-state gate with active HIGH enable.

The above figure shows when Enable input is HIGH, the NAND function provides 0 or 1 depending upon the input combination. If enable is LOW, the NAND gate output will provide high-impedance state independent of the input combinations.

Let us now consider NAND gate with the active LOW enable input.

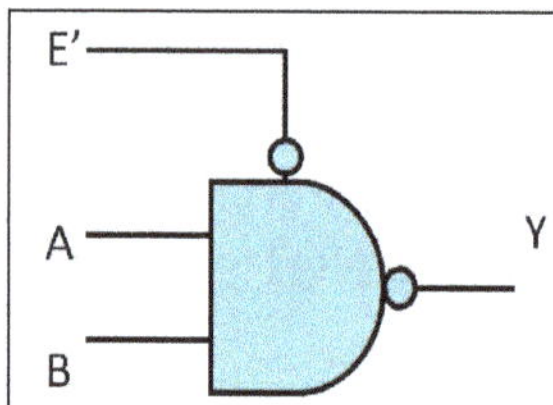

E	A	B	Y
0	0	0	1
0	0	1	1
0	1	0	1
0	1	1	0
1	0	0	Z
1	0	1	Z
1	1	0	Z
1	1	1	Z

Tri-state gate with active LOW enable.

The high-impedance state is controlled by an external ENABLE input. The ENABLE input decides whether the gate is active or in the high-impedance state. When active, it can be '0' or '1' depending upon input conditions. In this gate when Enable input is LOW, the NAND function provides 0 or 1 depending upon the input combination. If enable is HIGH, the NAND gate output will provide high-impedance state independent of the input combinations.

Let us consider the interconnections of tristate inverters together to form a common bus as shown in figure.

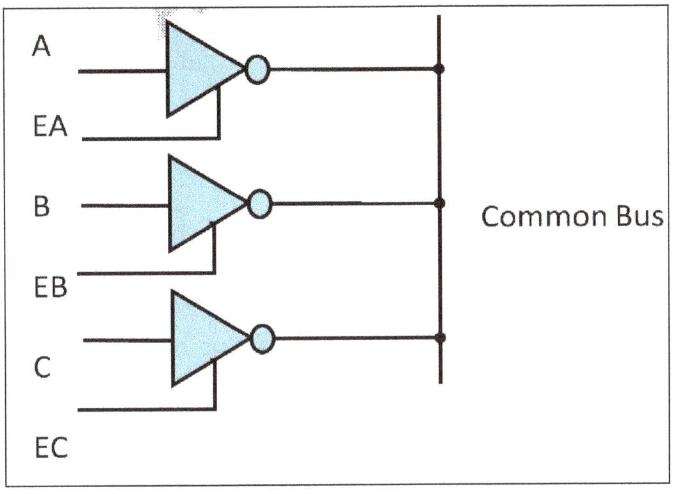

Tristate inverter.

All not gates are having single input and single output with enable input. The outputs are connected together to form a common bus. Each NOT gate has a separate active HIGH Enable input. This arrangement allows only one of them to be enabled at a time. HIGH ENABLE INPUT activates the respective NOT gate with a condition that only one of them is enabled at a time.

Schmitt Gates

A standard NOT gate or NAND gate is usually made up from transistor as a switching device. Transistor cannot switch from one state to another instantly; there is some delay in switching between the states. Transistor is basically a current amplifier and can also operate in linear region. Any small changes at its input level may cause a change in its output level and can even switch ON or OFF several times if there is any noise present in the system. One way to overcome this problem is to use Schmitt inverter (NOT or NAND gates).

Schmitt gate is basically a comparator with hysteresis. It is designed to operate when its input signal goes above an "Upper Threshold Point (UTP)", the output changes and goes to LOW level and remain in the same state until the input signal falls below the "Lower Threshold Point (LTP)". Once again the output changes and goes HIGH. In short, a Schmitt gate has built in Hysteresis to overcome the problem of noise.

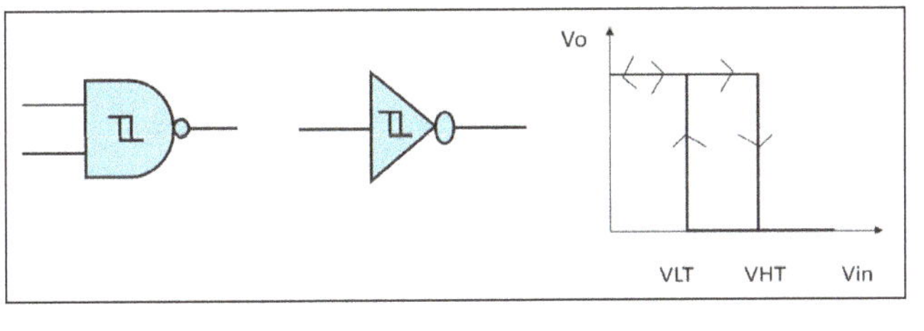

Figure shows NAND and NOT gate with built-in hysteresis. These gates are referred to as Schmitt gates. This gate interprets slowly changing input voltages according to two threshold voltages. Lower & upper threshold voltages are normally used for LOW to HIGH and HIGH to LOW transitions respectively. Schmitt gates are distinguished from conventional gates by the small 'hysteresis' symbol of the B–H loop for a ferromagnetic material. Hysteresis increases noise immunity of the digital system.

This clean ON/OFF switching makes the Schmitt gate ideal for switching any slow-rising or slow falling input signal. An important application of Schmitt gate is in oscillators i.e. sine to square wave converter or a clock.

Buffers

A buffer has only one input and one output. Normal logic gates have a limited load-driving capability. A logic gate can be treated as signal amplifiers, regardless of what logic function they may perform. For this purpose, a special logic gate called a buffer is designed. A buffer has a larger load-driving capability than a normal logic gate. It could be an inverting or non-inverting buffer with a single input, a single output and an Enable input. Figure below shows the symbols and functional tables of inverting and non-inverting buffers of the tri-state type. Buffer is also popularly known as "Driver". Buffers are usually tri-state devices preferred for designing bus-oriented system.

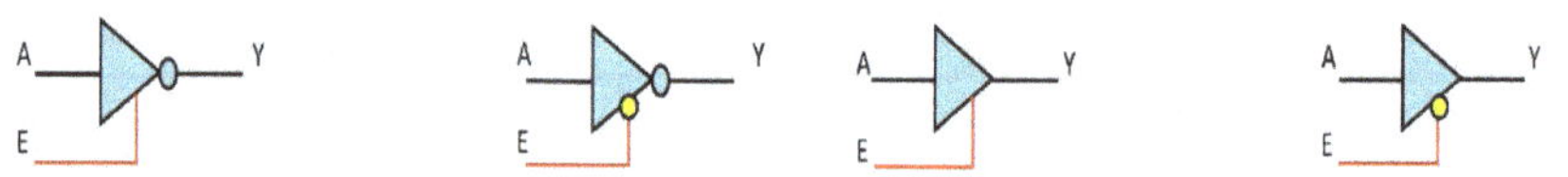

E	A	Y
0	X	Z
1	0	1
1	1	0

E'	A	Y
1	X	Z
0	0	1
0	1	0

E	A	Y
0	X	Z
1	0	0
1	1	1

E'	A	Y
1	X	Z
0	0	0
0	1	1

Inverting and non-inverting Buffers.

A tristate buffer is a useful device that allows us to control when current passes through the device and when it does not. A tri-state buffer has two inputs: a data input A and a enable input E. When the enable input is active i.e. At logic 1 then the output is same as input. If enable input is 0 then the output is in tri-state. In many applications we need tri-state buffers with active LOW enable input.

Transceivers

It is kind of buffer gate capable of both transmit and receive. A transceiver is a combination of transmitter and receiver in a single IC package. It is capable of driving bidirectional systems. A transceiver is a bidirectional buffer capable of driving bidirectional system with additional direction control/ENABLE inputs. It allows flow of data in both directions, depending upon the status of the control inputs. Transceivers can be popularly used to construct bus-oriented systems. Figure below indicates the circuit symbols of inverting and non-inverting transceivers.

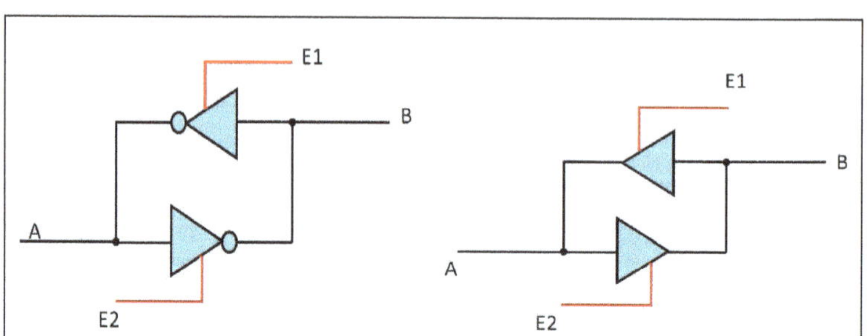

Inverting and non-inverting transceiver.

In case of inverting transceiver, when active HIGH input is applied at Enable input E1, data is transferred from A to B in complemented form and when Enable input E2=1, then the data is transferred from B to A in complemented form. Where as in non-inverting transceiver, data is transmitted or received in un-complemented form.

AOI Gate

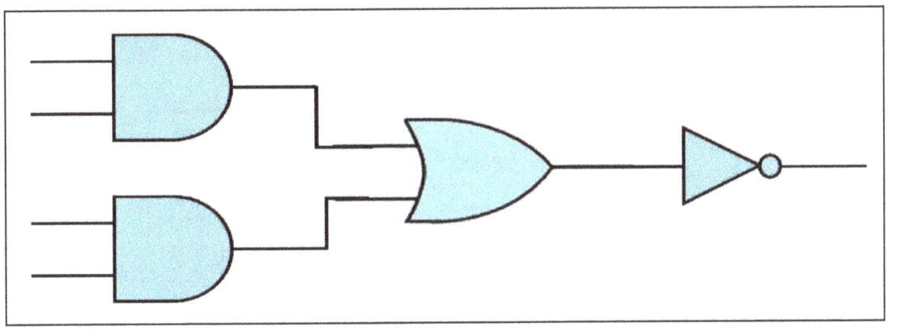

Figure indicates the special arrangement of three basic logic gates, which is popularly known as AOI (AND-OR-Invert) gate. It is used in many applications of combinational systems.

Pulsed Operations of Gates

In many applications, the input to a gate are not stationary levels but changing dynamically between two voltage levels frequently. These types of changing voltage or logic levels give rise to Pulse waveform. Let us now study the response of different logic gates with pulse input waveform.

Pulsed Operation of NOT Gate

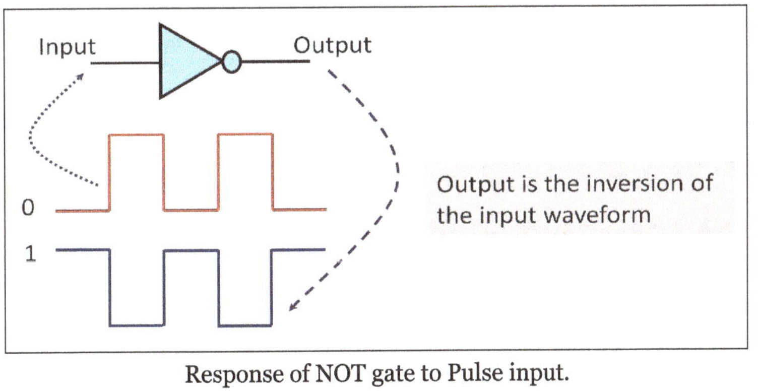

Response of NOT gate to Pulse input.

Let us now study the response of NOT gate to the pulse input waveform. It is important to note that the all logic gates obey the truth table irrespective of the static or dynamic inputs. For a NOT gate if the input is at logic 0 then output will be in logic 1 and vice a versa. The output is inverted as compared to input.

Pulsed Operation of AND Gate

In order to study the pulse response, it is important to look at the input of the gate with respect to each other to determine the output level at any time. Let us consider two AND gates with one of the input is fixed and tied to either 0 or 1 and pulse waveform is applied to the other input as shown in figure below. When one of the inputs is 1, the pulse appears as it at output and output is 0 when one of the inputs is 0.

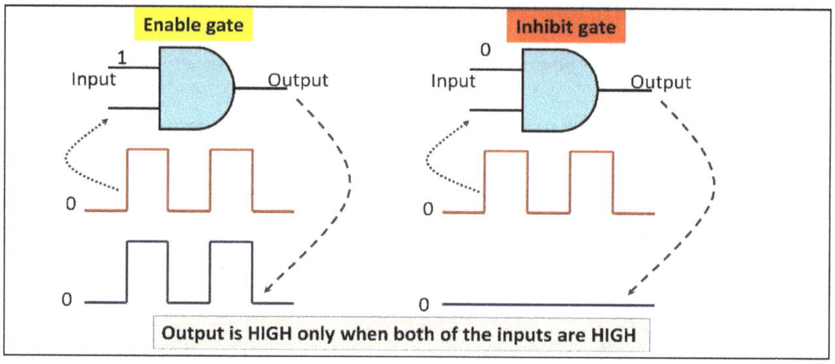

Response of AND gate to Pulse input.

Pulsed Operation of OR Gate

Let us consider two OR gates with one of the input is fixed and tied to either 0 or 1 and pulse waveform is applied to the other input as shown in figure below. When one of the inputs is 0, the pulse appears as it at output and output is 1 when one of the inputs is 1.

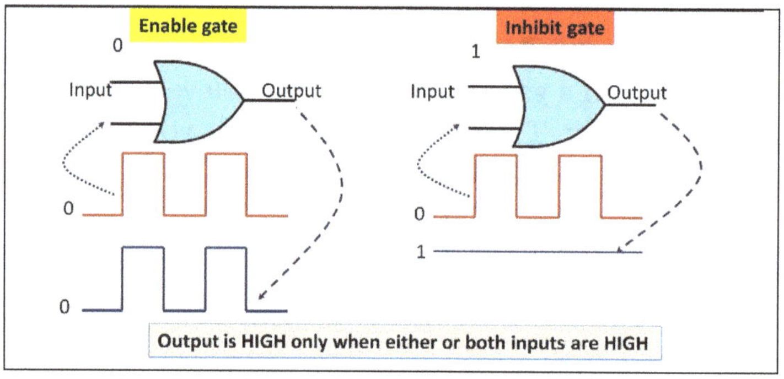

Response of OR gate to Pulse input.

Pulsed Operation of NAND Gate

Let us consider two NAND gates with one of the input is fixed and tied to either 0 or 1 and pulse waveform is applied to the other input as shown in figure below. When one of the inputs is 1, the pulse appears inverted at output and output is 1 when one of the inputs is 0.

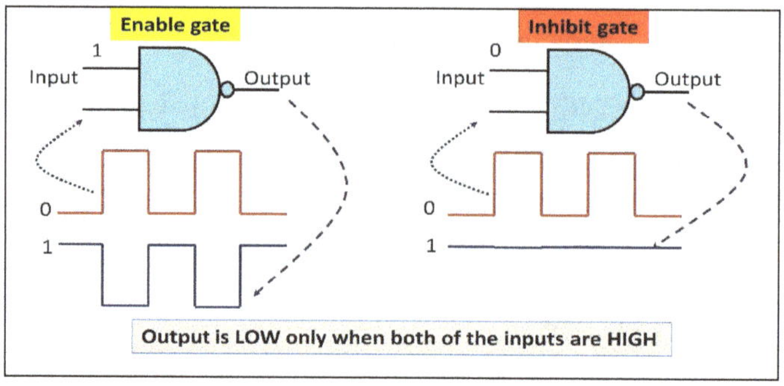

Response of NAND gate to Pulse input.

Pulsed Operation of NOR Gate

Let us consider two NOR gates with one of the input is fixed and tied to either 0 or 1 and pulse waveform is applied to the other input as shown in figure below. When one of the inputs is 0, the pulse inverted at output and output is 0 when one of the inputs is 1.

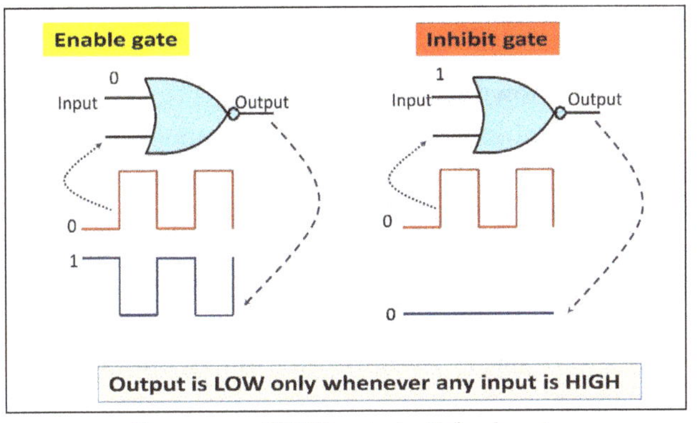

Response of NOR gate to Pulse input.

References

- Not-gate: circuitglobe.com, Retrieved 15, August 2020
- Logic-2: electronics-tutorials.ws, Retrieved 20, February 2020
- Logic-3: electronics-tutorials.ws, Retrieved 30, April 2020
- Nand-gate-in-digital-electronics: javatpoint.com, Retrieved 08, June 2020
- Nor-gate-in-digital-electronics: javatpoint.com, Retrieved 17, January 2020

4

Number Systems: An Integrated Study

The system of naming numbers is known as number system. The different bases of number system in digital electronics are the decimal, binary, octal, and hexadecimal number system. The chapter discusses the key types of number systems such as binary number system, binary code, binary-coded decimal, bitwise operation and Boolean algebra to provide an extensive understanding of the subject.

Any symbol can be represented as a number and the set of symbols and digits to represent information in a code system can be expressed as a set of mathematical numbers. Every number in a code system or encoding scheme refers to a unique symbol or digit and can be expressed in binary form i.e. in the form of bits or bytes.

There are many number systems in use, commonly decimal, octal and hexadecimal. The numbers in any number system can be converted to binary system and so represented in binary form, and so in digital electronic form.

Most number systems follow a common pattern for writing down the value of a number.

A fixed number of values can be written with a single numerical character, then a new column is used to count how many times the highest value in the counting system has been reached. The number of numerical values the system uses is called the base of the system. For example, the decimal system has 10 numerical characters and so has a base of 10:

0 1 2 3 4 5 6 7 8 9

For writing numbers greater than 9 a second column is added to the left, and this column has 10 times the value of the column immediately to its right.

Because number systems commonly used in digital electronics have different base values to the decimal system, they look less familiar, but work in essentially the same way.

Decimal, (Base 10)

Decimal has ten values 0 to 9. If larger values than 9 are needed, extra columns are added to the left. Each column value is ten times the value of the column to its right. For example the decimal value twenty-two is written 22 (2 tens + 2 ones).

Binary, (Base 2)

Binary has only two values 0 and 1. If larger values than 1 are needed, extra columns are added to the left. Each column value is now twice the value of the column to its right. For example the decimal value three is written 11 in binary (1 two + 1 one).

Octal, (Base 8)

Octal has eight values 0 to 7. If larger values than 7 are needed, extra columns are added to the left. Each column value is now 8 times the value of the column to its right. For example the decimal value twenty-seven is written 33 in octal (3 eights + 3 ones).

Hexadecimal, (Base 16)

Hexadecimal has sixteen values 0 to 15, but to keep all these values in a single column, the 16 values (0 to 15) are written as 0 to F, using the letters A to F to represent numbers 10 to 15, so avoiding the use of a second column. Again, if higher values than 15 (F in hexadecimal) are needed, extra columns to the left are used. Each column value is sixteen times that of the column to its right. For example the decimal value sixty-eight is written as 44 in hexadecimal (4 sixteens + 4 ones).

Some column values of different number systems				
Decimal	1000	100	10	1
Binary	8	4	2	1
Octal	512	64	8	1
Hexadecimal	4096	256	16	1

Each of these different number systems works in the same way, it is just that each system has a different base, and the column values in each system increase by multiples of the base number as columns are added to the left.

Because this topic describes several different number systems, it is important to know which system is being described. Therefore if there is some doubt which system a number is in, the base of the system, written as a subscript immediately after the value, is used to identify the number system.

For example:

10_{10} represents the decimal value ten. (1 ten + 0 units)

10_{2} represents the binary value two. (1 two + 0 units)

10_{8} represents the octal value eight. (1 eight + 0 units)

10_{16} represents the hexadecimal value sixteen. (1 sixteen + 0 units)

System Radix

The base of a system, more properly called the RADIX, is the number of different values that can be expressed using a single digit. Therefore the decimal system has a radix of 10, the octal system has a radix of 8, hexadecimal is radix 16, and binary radix 2.

Decimal (Radix 10)	Binary (Radix 2)	Octal (Radix 8)	Hexadecimal (Radix 16)
0	0	0	0
1	1	1	1
2		2	2
3		3	3
4		4	4
5		5	5
6		6	6
7		7	7
8			8
9			9
			A
			B
			C
			D
			E
			F

The range of number values in different number systems is shown in table above, Notice that because the hexadecimal system must express 16 values using only one column, it uses the letters A B C D E & F to represent the numbers 10 to 15.

Radix Point

When writing a number, the digits used give its value, but the number is 'scaled' by its RADIX POINT.

For example, 456.2_{10} is ten times bigger than 45.62_{10} although the digits are the same.

Notice also that when using multiple number systems, the term 'RADIX point' instead of 'DECIMAL point' is used. When using decimal numbers, a decimal point is used, but if a different system is used, it would be wrong to call the point a decimal point, it would need to be called "Binary point" or "Octal point" etc. The simplest way around this is to refer to the point in any system (which will of course have its value labelled with its radix) as the RADIX POINT.

Exponents

A decimal number such as 456.2_{10} can be considered as the sum of the values of its individual digits, where each digit has a value dependent on its position within the number (the value of the column):

$= 456.2_{10}$

Col 2	Col 1	Col 0	Col -1
4 hundreds	+ 5 tens	+ 6 units	+ 2 tenths
(4×10^2)	$+ (5 \times 10^1)$	$+ (6 \times 10^0)$	$+ (2 \times 10^{-1})$
400	+ 50	+ 6	+ 0.2

Each digit in the number is multiplied by the system radix raised to a power depending on its position relative to the radix point. This is called the EXPONENT. The digit immediately to the left of the radix point has the exponent 0 applied to its radix, and for each place to the left, the exponent increases by one. The first place to the right of the radix point has the exponent -1 and so on, positive exponents to the left of the radix point and negative exponents to the right.

This method of writing numbers is widely used in electronics with decimal numbers, but can be used with any number system. Only the radix is different.

Hexadecimal exponents $98.2_{16} = (9 \times 16^1) + (8 \times 16^0) + (2 \times 16^{-1})$

Octal exponents $56.2_8 = (5 \times 8^1) + (6 \times 8^0) + (2 \times 8^{-1})$

Binary Exponents $10.1_2 = (1 \times 2^1) + (0 \times 2^0) + (1 \times 2^{-1})$

When using your calculator for the above examples you may find that it does not like radix points in anything other than decimal mode. This is common with many electronic calculators.

Floating Point Notation

If electronic calculators cannot use radix points other than in decimal, this could be a problem. Fortunately for every problem there is a solution. The radix exponent can also be used to eliminate the radix point, without altering the value of the number. In the example below, see how the value remains the same while the radix point moves. It is all done by changing the radix exponent.

$102.6_{10} = 102.6 \times 10^0 = 10.26 \times 10^1 = 1.026 \times 10^2 = .1026 \times 10^3$

The radix point is moved one place to the left by increasing the exponent by one.

It is also possible to move the radix point to the right by decreasing the exponent. In this way the radix point can be positioned wherever it is required, in any number system, simply by changing the exponent. This is called FLOATING POINT NOTATION and it is how calculators handle decimal points in calculations.

Normalised Form

By putting the radix point at the front of the number, and keeping it there by changing the exponent, calculations become easier to do electronically, in any radix.

Electronic Storage of Numbers

A number written (or stored) in this way, with the radix point at the left of the most significant digit is said to be in NORMALISED FORM. For example $.11011_2 \times 2^3$ is the normalised form of the binary number 110.11_2. Because numbers in electronic systems are stored as binary digits, and a binary digit can only be 1 or 0, it is not possible to store the radix point within the number. Therefore the number is stored in its normalised form and the exponent is stored separately. The exponent is then reused to restore the radix point to its correct position when the number is displayed.

In electronics systems a single binary digit is called a bit (short for Binary DigIT), but as using a single digit would seriously limit the maths that could be performed, binary bits are normally used in groups.

4 bits = 1 nibble

8 bits = 1 byte

Multiple bytes, such as 16 bits, 32 bits, 64 bits are usually called 'words', e.g. a 32 bit word. The length of the word depends on how many bits can be physically handled or stored by the system at one time.

4 Bit Binary Representation

Decimal	MSB	4 Bit Binary		LSB
	$2^3 = 8$	$2^2 = 4$	$2^1 = 2$	$2^0 = 1$
0	0	0	0	0
1	0	0	0	1
2	0	0	1	0
3	0	0	1	1
4	0	1	0	0
5	0	1	0	1
6	0	1	1	0
7	0	1	1	1
8	1	0	0	0
9	1	0	0	1
10	1	0	1	0
11	1	0	1	1
12	1	1	0	0
13	1	1	0	1
14	1	1	1	0
15	1	1	1	1

When a number is stored in an electronic system, it is stored in a memory location having a fixed number of binary bits. Some of these memory locations are used for general storage whilst others, having some special function, are called registers. Wherever a number is stored, it will be held in some form of binary, and must always have a set number of bits. Therefore a decimal number such as 13, which can be expressed in four binary bits as 1101_2 becomes 00001101_2 when stored in an eight-bit register. This is achieved by adding four NON SIGNIFICANT ZEROS to the left of the most significant '1' digit.

Using this system, a binary register that is n bits wide can hold 2^n values.

Therefore an 8 bit register can hold 2^8 values = 256 values (0 to 255).

A 4 bit register can hold 2^4 values = 16 values (0 to 15).

Binary Number System

Binary numbers are the flow of information in the form of zeros and ones used by digital computers and systems.

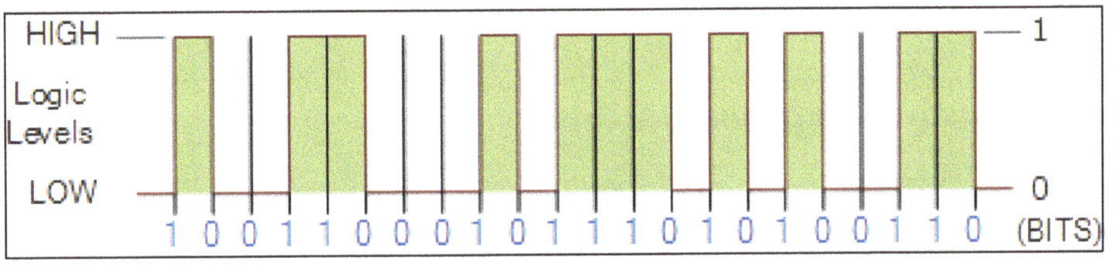

Unlike a linear, or analogue circuits, such as AC amplifiers, which process signals that are constantly changing from one value to another, for example amplitude or frequency, digital circuits process signals that contain just two voltage levels or states, labelled, Logic "0" and Logic "1".

Generally, a logic "1" represents a higher voltage, such as 5 volts, which is commonly referred to as a HIGH value, while a logic "0" represents a low voltage, such as 0 volts or ground, and is commonly referred to as a LOW value. These two discrete voltage levels representing the digital values of "1's" (one's) and "0's" (zero's) are commonly called: Binary digits, and in digital and computational circuits and applications they are normally referred to as binary BITS.

Binary Bits of Zeros and Ones

Because there are only two valid Boolean values for representing either a logic "1" or a logic "0", makes the system of using Binary Numbers ideal for use in digital or electronic circuits and systems.

The *binary number system* is a Base-2 numbering system which follows the same set of rules in mathematics as the commonly used decimal or base-10 number system. So instead of powers of ten, (10^n) for example: 1, 10, 100, 1000 etc, binary numbers use powers of two, (2^n) effectively doubling the value of each successive bit as it goes, for example: 1, 2, 4, 8, 16, 32 etc.

The voltages used to represent a digital circuit can be of any value, but generally in digital and computer systems they are kept well below 10 volts. In digital systems theses voltages are called "logic levels" and ideally one voltage level represents a "HIGH" state, while another different and lower voltage level represents a "LOW" state. A binary number system uses both of these two states.

Digital waveforms or signals consist of discrete or distinctive voltage levels that are changing back and forth between these two "HIGH" and "LOW" states. But what makes a signal or voltage "Digital" and how can we represent these "HIGH" and "LOW" voltage levels.

Digital Voltage Output

In this digital circuit example, the potentiometer wiper has been replaced by a single rotary switch which is connected in turn to each junction of the series resistor chain, forming a basic potential divider network. As the switch is rotated from one position (or node) to the next the output voltage, V_{out} changes quickly in discrete and distinctive voltage levels representing multiples of 1.0 volts on each switching action or step, as shown in the output graph.

So for example, the output voltage will be 2 volts, 3 volts, 5 volts, etc. but NOT 2.5V, 3.1V or 4.6V. Finer output voltage levels could easily be produced by using a multi-positional switch and increasing the number of resistive elements within the potential divider network, therefore increasing the number of discrete switching steps.

Digital Voltage Output Representation

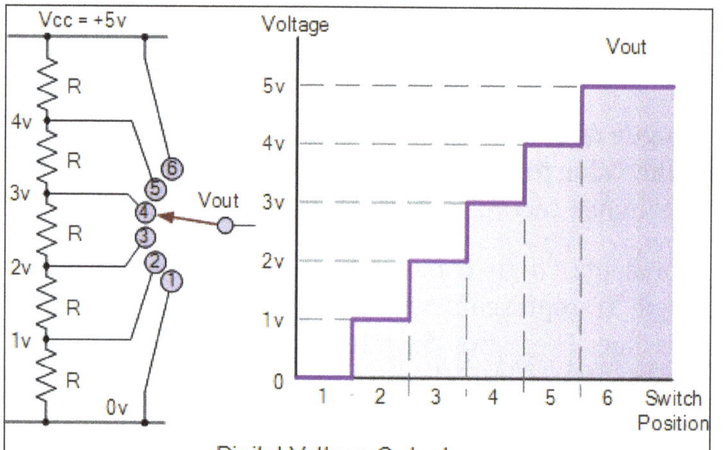

Digital Voltage Output

Then we can see that the major difference between an analogue signal or quantity and a digital quantity is that an "Analogue" quantity is continuously changing over time while a "Digital" quantity has discrete (step by step) values. "LOW" to "HIGH" or "HIGH" to "LOW".

A good example of this would be a light dimmer in your house that varies the lights intensity (brightness) up or down as it is rotated between fully-ON (maximum brightness) and fully-OFF, producing an analogue output that varies continuously. While on the other hand, with a standard wall mounted light switch, the light is either "ON" (HIGH) or it is "OFF" (LOW) when the switch is operated. The result is that there is no in between producing a form of ON-OFF digital output.

Some circuits combine both analogue and digital signals such as an analogue to digital converter (ADC) or a digital to analogue converter (DAC). Either way, the digital input or output signal represents a binary number value equivalent of an analogue signal.

Digital Value Representation

First State	Second State
Logic “0”	Logic “1”
LOW	HIGH
FALSE	TRUE
Low Level Voltage Output	High Level Voltage Output
0V or Ground	+5 Volts
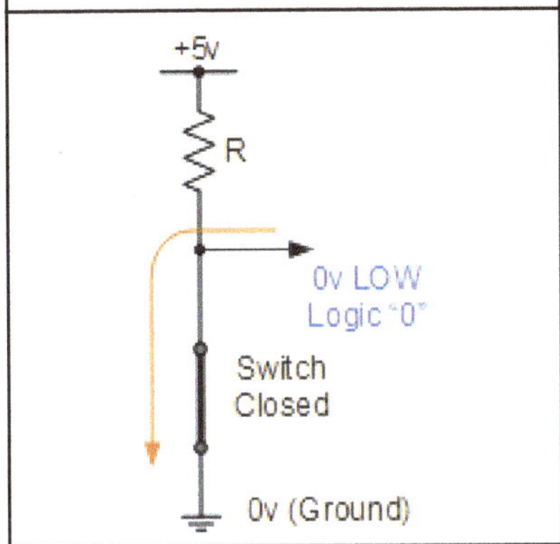	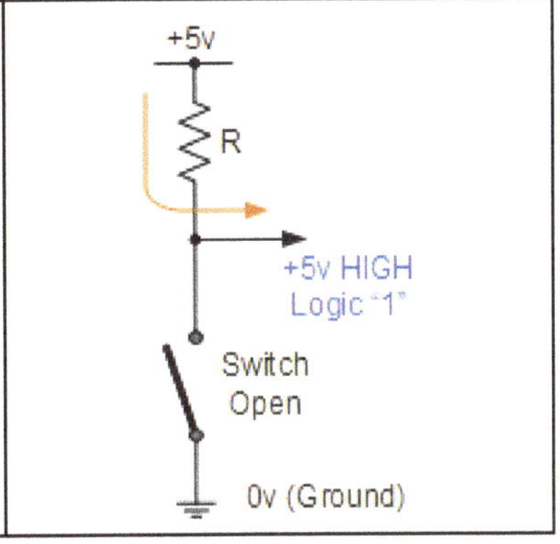

Generally the switching from one voltage level, “>0” to “1” or “1” to “0” is made as quickly as possible to prevent miss switching of the logic circuit. In standard TTL (transistor-transistor-logic) IC’s there is a pre-defined range of input and output voltage limits for defining what exactly is a logic “1” value and what is a logic “0” value as shown below.

Representation

Any number can be represented by a sequence of bits (binary digits), which in turn may be represented by any mechanism capable of being in two mutually exclusive states. Any of the following rows of symbols can be interpreted as the binary numeric value of 667:

1	0	1	0	0	1	1	0	1	1
|	–	|	–	–	|	|	–	|	|
☒	☐	☒	☐	☐	☒	☒	☐	☒	☒
y	n	y	n	n	y	y	n	y	y

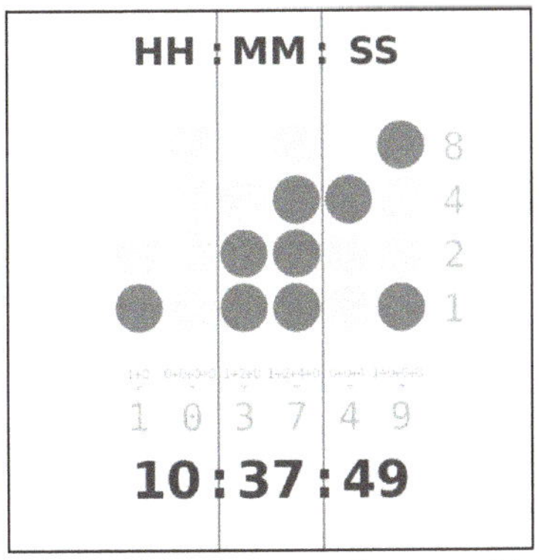

A binary clock might use LEDs to express binary values. In this clock, each column of LEDs shows a binary-coded decimal numeral of the traditional sexagesimal time.

The numeric value represented in each case is dependent upon the value assigned to each symbol. In a computer, the numeric values may be represented by two different voltages; on a magnetic disk, magnetic polarities may be used. A "positive", "yes", or "on" state is not necessarily equivalent to the numerical value of one; it depends on the architecture in use.

In keeping with customary representation of numerals using Arabic numerals, binary numbers are commonly written using the symbols 0 and 1. When written, binary numerals are often subscripted, prefixed or suffixed in order to indicate their base, or radix. The following notations are equivalent:

- 100101 binary (explicit statement of format)
- 100101b (a suffix indicating binary format; also known as Intel convention)
- 100101B (a suffix indicating binary format)
- bin 100101 (a prefix indicating binary format)
- 100101_2 (a subscript indicating base-2 (binary) notation)
- %100101 (a prefix indicating binary format; also known as Motorola convention)
- 0b100101 (a prefix indicating binary format, common in programming languages)
- 6b100101 (a prefix indicating number of bits in binary format, common in programming languages)

When spoken, binary numerals are usually read digit-by-digit, in order to distinguish them from decimal numerals. For example, the binary numeral 100 is pronounced *one zero zero*, rather than *one hundred*, to make its binary nature explicit, and for purposes of correctness. Since the binary numeral 100 represents the value four, it would be confusing to refer to the numeral as *one hundred* (a word that represents a completely different value, or amount). Alternatively, the binary numeral 100 can be read out as "four" (the correct *value*), but this does not make its binary nature explicit.

Counting in Binary

Decimal pattern	Binary number
0	0
1	1
2	10
3	11
4	100
5	101
6	110
7	111
8	1000

9	1001
10	1010
11	1011
12	1100
13	1101
14	1110
15	1111

Counting in binary is similar to counting in any other number system. Beginning with a single digit, counting proceeds through each symbol, in increasing order. Before examining binary counting, it is useful to briefly discuss the more familiar decimal counting system as a frame of reference.

Decimal Counting

Decimal counting uses the ten symbols *0* through *9*. Counting begins with the incremental substitution of the least significant digit (rightmost digit) which is often called the *first digit*. When the available symbols for this position are exhausted, the least significant digit is reset to *0*, and the next digit of higher significance (one position to the left) is incremented (*overflow*), and incremental substitution of the low-order digit resumes. This method of reset and overflow is repeated for each digit of significance. Counting progresses as follows:

000, 001, 002, 007, 008, 009, (rightmost digit is reset to zero, and the digit to its left is incremented)

010, 011, 012

090, 091, 092, 097, 098, 099, (rightmost two digits are reset to zeroes, and next digit is incremented)

100, 101, 102

Binary Counting

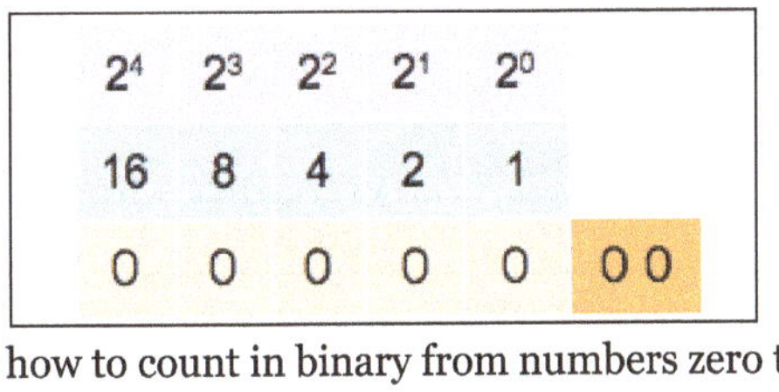

This counter shows how to count in binary from numbers zero through thirty-one.

Binary counting follows the same procedure, except that only the two symbols *0* and *1* are available. Thus, after a digit reaches 1 in binary, an increment resets it to 0 but also causes an increment of the next digit to the left:

0000,

0001, (rightmost digit starts over, and next digit is incremented)

0010, 0011, (rightmost two digits start over, and next digit is incremented)

0100, 0101, 0110, 0111, (rightmost three digits start over, and the next digit is incremented)

1000, 1001, 1010, 1011, 1100, 1101, 1110, 1111

In the binary system, each digit represents an increasing power of 2, with the rightmost digit representing 2^0, the next representing 2^1, then 2^2, and so on. The equivalent decimal representation of a binary number is sum of the powers of 2 which each digit represents. For example, the binary number 100101 is converted to decimal form as follows:

$$100101_2 = [(1) \times 2^5] + [(0) \times 2^4] + [(0) \times 2^3] + [(1) \times 2^2] + [(0) \times 2^1] + [(1) \times 2^0]$$

$$100101_2 = [1 \times 32] + [0 \times 16] + [0 \times 8] + [1 \times 4] + [0 \times 2] + [1 \times 1]$$

$$100101_2 = 37_{10}$$

Fractions

Fractions in binary only terminate if the denominator has 2 as the only prime factor. As a result, 1/10 does not have a finite binary representation, and this causes 10 × 0.1 not to be precisely equal to 1 in floating point arithmetic. As an example, to interpret the binary expression for 1/3 = .010101..., this means: $1/3 = 0 \times 2^{-1} + 1 \times 2^{-2} + 0 \times 2^{-3} + 1 \times 2^{-4} + ... = 0.3125 + ...$ An exact value cannot be found with a sum of a finite number of inverse powers of two, the zeros and ones in the binary representation of 1/3 alternate forever.

Fraction	Decimal	Binary	Fractional approximation
1/1	1 or 0.999...	1 or 0.111...	1/2 + 1/4 + 1/8...
1/2	0.5 or 0.4999...	0.1 or 0.0111...	1/4 + 1/8 + 1/16 . . .
1/3	0.333...	0.010101...	1/4 + 1/16 + 1/64 . . .
1/4	0.25 or 0.24999...	0.01 or 0.00111...	1/8 + 1/16 + 1/32 . . .
1/5	0.2 or 0.1999...	0.00110011...	1/8 + 1/16 + 1/128 . . .
1/6	0.1666...	0.0010101...	1/8 + 1/32 + 1/128 . . .
1/7	0.142857142857...	0.001001...	1/8 + 1/64 + 1/512 . . .
1/8	0.125 or 0.124999...	0.001 or 0.000111...	1/16 + 1/32 + 1/64 . . .
1/9	0.111...	0.000111000111...	1/16 + 1/32 + 1/64 . . .
1/10	0.1 or 0.0999...	0.000110011...	1/16 + 1/32 + 1/256 . . .
1/11	0.090909...	0.00010111010001011101...	1/16 + 1/64 + 1/128 . . .
1/12	0.08333...	0.00010101...	1/16 + 1/64 + 1/256 . . .
1/13	0.076923076923...	0.000100111011000100111011...	1/16 + 1/128 + 1/256 . . .
1/14	0.0714285714285...	0.0001001001...	1/16 + 1/128 + 1/1024 . . .
1/15	0.0666...	0.00010001...	1/16 + 1/256 . . .
1/16	0.0625 or 0.0624999...	0.0001 or 0.0000111...	1/32 + 1/64 + 1/128 . . .

Binary Arithmetic

Arithmetic in binary is much like arithmetic in other numeral systems. Addition, subtraction, multiplication, and division can be performed on binary numerals.

Addition

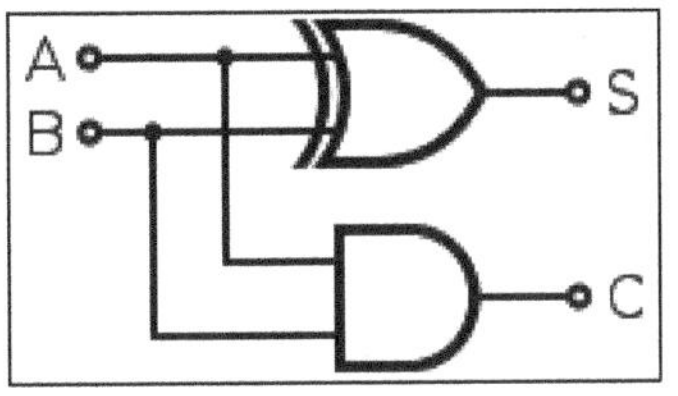

The circuit diagram for a binary half adder, which adds two bits together, producing sum and carry bits.

The simplest arithmetic operation in binary is addition. Adding two single-digit binary numbers is relatively simple, using a form of carrying:

$0 + 0 \rightarrow 0$

$0 + 1 \rightarrow 1$

$1 + 0 \rightarrow 1$

$1 + 1 \rightarrow 0$, carry 1 (since $1 + 1 = 2 = 0 + (1 \times 2^1)$)

Adding two "1" digits produces a digit "0", while 1 will have to be added to the next column. This is similar to what happens in decimal when certain single-digit numbers are added together; if the result equals or exceeds the value of the radix (10), the digit to the left is incremented:

$5 + 5 \rightarrow 0$, carry 1 (since $5 + 5 = 10 = 0 + (1 \times 10^1)$)

$7 + 9 \rightarrow 6$, carry 1 (since $7 + 9 = 16 = 6 + (1 \times 10^1)$)

This is known as *carrying*. When the result of an addition exceeds the value of a digit, the procedure is to "carry" the excess amount divided by the radix (that is, 10/10) to the left, adding it to the next positional value. This is correct since the next position has a weight that is higher by a factor equal to the radix. Carrying works the same way in binary:

```
  1 1 1 1 1     (carried digits)
   0 1 1 0 1
 +  1 0 1 1 1
-------------
= 1 0 0 1 0 0 = 36
```

In this example, two numerals are being added together: 01101_2 (13_{10}) and 10111_2 (23_{10}). The top row shows the carry bits used. Starting in the rightmost column, $1 + 1 = 10_2$. The 1 is carried to the left, and the 0 is written at the bottom of the rightmost column. The second column from the right is added: $1 + 0 + 1 = 10_2$ again; the 1 is carried, and 0 is written at the bottom. The third column: $1 + 1 + 1 = 11_2$. This time, a 1 is carried, and a 1 is written in the bottom row. Proceeding like this gives the final answer 100100_2 (36 decimal).

When computers must add two numbers, the rule that: x xor y = (x + y) mod 2 for any two bits x and y allows for very fast calculation, as well.

Long Carry Method

A simplification for many binary addition problems is the Long Carry Method or Brookhouse Method of Binary Addition. This method is generally useful in any binary addition where one of the numbers contains a long "string" of ones. It is based on the simple premise that under the binary system, when given a "string" of digits composed entirely of n ones (*where:* n is any integer length), adding 1 will result in the number 1 followed by a string of n zeros. That concept follows, logically, just as in the decimal system, where adding 1 to a string of n 9s will result in the number 1 followed by a string of n 0s:

```
      Binary                          Decimal
       11111         likewise          99999
+          1                          +    1
 ------                              --------
 100000                               100000
```

Such long strings are quite common in the binary system. From that one finds that large binary numbers can be added using two simple steps, without excessive carry operations. In the following example, two numerals are being added together: $1\,1\,1\,0\,1\,1\,1\,1\,1\,0_2$ (958_{10}) and $1\,0\,1\,0\,1\,1\,0\,0\,1\,1_2$ (691_{10}), using the traditional carry method on the left, and the long carry method on the right:

Traditional Carry Method vs. Long Carry Method

111 11111 (carried digits) 1← 1← carry the 1 until it is one digit past the "string" below

111 0111110 ~~111~~ 0 ~~11111~~ 0 cross out the "string",

+1010110011 + 10 ~~1~~ 01100 ~~1~~ 1 and cross out the digit that was added to it

---

=1100111 0001 1100111 0001

The top row shows the carry bits used. Instead of the standard carry from one column to the next, the lowest-ordered "1" with a "1" in the corresponding place value beneath it may be added and a "1" may be carried to one digit past the end of the series. The "used" numbers must be crossed off, since they are already added. Other long strings may likewise be cancelled using the same technique. Then, simply add together any remaining digits normally. Proceeding in this manner gives the final answer of $1\,1\,0\,0\,1\,1\,1\,0\,0\,0\,1_2$ (1649_{10}). In our simple example using small numbers, the traditional carry method required eight carry operations, yet the long carry method required only two, representing a substantial reduction of effort.

Addition Table

	0	1
0	0	1
1	1	10

The binary addition table is similar, but not the same, as the truth table of the logical disjunction operation $\vee$. The difference is that $\vee 1 = 1$, while $1 + 1 = 10$.

Subtraction

Subtraction works in much the same way:

$0 - 0 \rightarrow 0$

$0 - 1 \rightarrow 1$, borrow 1

$1 - 0 \rightarrow 1$

$1 - 1 \rightarrow 0$

Subtracting a "1" digit from a "0" digit produces the digit "1", while 1 will have to be subtracted from the next column. This is known as *borrowing*. The principle is the same as for carrying. When the result of a subtraction is less than 0, the least possible value of a digit, the procedure is to "borrow" the deficit divided by the radix (that is, 10/10) from the left, subtracting it from the next positional value.

```
  *   * * *  (starred columns are borrowed from)
 1 1 0 1 1 1 0
 -     1 0 1 1 1
----------------
=1 0 1 0 1 1 1

  *         (starred columns are borrowed from)
 1 0 1 1 1 1 1
 -   1 0 1 0 1 1
----------------
=0 1 1 0 1 0 0
```

Subtracting a positive number is equivalent to *adding* a negative number of equal absolute value. Computers use signed number representations to handle negative numbers—most commonly the two's complement notation. Such representations eliminate the need for a separate "subtract" operation. Using two's complement notation subtraction can be summarized by the following formula:

$A - B = A + \text{not } B + 1$

Multiplication

Multiplication in binary is similar to its decimal counterpart. Two numbers A and B can be multiplied by partial products: for each digit in B, the product of that digit in A is calculated and written on a new line, shifted leftward so that its rightmost digit lines up with the digit in B that was used. The sum of all these partial products gives the final result.

Since there are only two digits in binary, there are only two possible outcomes of each partial multiplication:

- If the digit in B is 0, the partial product is also 0

- If the digit in B is 1, the partial product is equal to A

For example, the binary numbers 1011 and 1010 are multiplied as follows:

```
        1011  (A)
      ×1010  (B)
      ---------
        0000   ← Corresponds to the rightmost 'zero' in B
    + 1011     ← Corresponds to the next 'one' in B
   + 0000
  +1011
  ---------------
  =1101110
```

Binary numbers can also be multiplied with bits after a binary point:

```
          101.101       A (5.625 in decimal)
         ×110.01        B (6.25 in decimal)
         -------------------
              1.01101  ← Corresponds to a 'one' in B
  +         00.0000    ← Corresponds to a 'zero' in B
  +       000.000
  +      1011.01
  +     10110.1
  ---------------------------
  =   100011.00101 (35.15625 in decimal)
```

Multiplication Table

	0	1
0	0	0
1	0	1

The binary multiplication table is the same as the truth table of the logical conjunction operation .

Division

Long division in binary is again similar to its decimal counterpart.

In the example below, the divisor is 101_2, or 5 decimal, while the dividend is 11011_2, or 27 decimal. The procedure is the same as that of decimal long division; here, the divisor 101_2 goes into the first three digits 110_2 of the dividend one time, so a "1" is written on the top line. This result is multiplied by the divisor, and subtracted from the first three digits of the dividend; the next digit (a "1") is included to obtain a new three-digit sequence:

```
          1
     ____________
101 )11011
    -101
     -----
     001
```

The procedure is then repeated with the new sequence, continuing until the digits in the dividend have been exhausted:

```
          101
     ____________
101 )11011
    -101
     ------
       111
    -  101
      ------
        10
```

Thus, the quotient of 11011_2 divided by 101_2 is 101_2, as shown on the top line, while the remainder, shown on the bottom line, is 10_2. In decimal, 27 divided by 5 is 5, with a remainder of 2.

Square Root

The process of taking a binary square root digit by digit is the same as for a decimal square root, and is explained here. An example is:

```
          1001
        ---------
        √1010001
         1
        ---------
101      01
          0
        --------
1001     100
           0
         --------
10001    10001
         10001
         -------
             0
```

Conversion to and from other Numeral Systems

Decimal

To convert from a base-10 integer to its base-2 (binary) equivalent, the number is divided by two. The remainder is the least-significant bit. The quotient is again divided by two; its remainder becomes the next least significant bit. This process repeats until a quotient of one is reached. The sequence of remainders (including the final quotient of one) forms the binary value, as each remainder must be either zero or one when dividing by two. For example, $(357)_{10}$ is expressed as $(101100101)_2$.

Conversion from base-2 to base-10 simply inverts the preceding algorithm. The bits o.f the binary number are used one by one, starting with the most significant (leftmost) bit. Beginning with the value 0, the prior value is doubled, and the next bit is then added to produce the next value. This can be organized in a multi-column table. For example, to convert 10010101101_2 to decimal:

Prior value	× 2 +	Next bit	Next value
0	× 2 +	1	= 1
1	× 2 +	0	= 2
2	× 2 +	0	= 4
4	× 2 +	1	= 9
9	× 2 +	0	= 18
18	× 2 +	1	= 37
37	× 2 +	0	= 74
74	× 2 +	1	= 149
149	× 2 +	1	= 299
299	× 2 +	0	= 598
598	× 2 +	1	= 1197

The result is 1197_{10}. Note that the first Prior Value of 0 is simply an initial decimal value. This method is an application of the Horner scheme.

Binary	1	0	0	1	0	1	0	1	1	0	1	
Decimal	1×2^{10} +	0×2^9 +	0×2^8 +	1×2^7 +	0×2^6 +	1×2^5 +	0×2^4 +	1×2^3 +	1×2^2 +	0×2^1 +	1×2^0 =	1197

The fractional parts of a number are converted with similar methods. They are again based on the equivalence of shifting with doubling or halving.

In a fractional binary number such as 0.11010110101_2, the first digit is $\frac{1}{2}$, the second $(\frac{1}{2})^2 = \frac{1}{4}$, etc. So if there is a 1 in the first place after the decimal, then the number is at least $\frac{1}{2}$, and vice versa. Double that number is at least 1. This suggests the algorithm: Repeatedly double the number to be converted, record if the result is at least 1, and then throw away the integer part.

For example, $(\frac{1}{3})_{10}$, in binary, is:

Converting	Result
$(\frac{1}{3})$	0.
$\frac{1}{3}\times 2=\frac{2}{3}<1$	0.0
$\frac{2}{3}\times 2=1\frac{1}{3}\geq 1$	0.01
$\frac{1}{3}\times 2=\frac{2}{3}<1$	0.010
$\frac{2}{3}\times 2=1\frac{1}{3}\geq 1$	0.0101

Thus the repeating decimal fraction $0.\overline{3}$... is equivalent to the repeating binary fraction 0.01... .

Or for example, 0.1_{10}, in binary, is:

Converting	Result
0.1	0.
0.1 × 2 = 0.2 < 1	0.0
0.2 × 2 = 0.4 < 1	0.00
0.4 × 2 = 0.8 < 1	0.000
0.8 × 2 = 1.6 ≥ 1	0.0001
0.6 × 2 = 1.2 ≥ 1	0.00011
0.2 × 2 = 0.4 < 1	0.000110
0.4 × 2 = 0.8 < 1	0.0001100
0.8 × 2 = 1.6 ≥ 1	0.00011001
0.6 × 2 = 1.2 ≥ 1	0.000110011
0.2 × 2 = 0.4 < 1	0.0001100110

This is also a repeating binary fraction $0.0\overline{0011}$... . It may come as a surprise that terminating decimal fractions can have repeating expansions in binary. It is for this reason that many are surprised to discover that 0.1 + ... + 0.1, (10 additions) differs from 1 in floating point arithmetic. In fact, the only binary fractions with terminating expansions are of the form of an integer divided by a power of 2, which 1/10 is not.

The final conversion is from binary to decimal fractions. The only difficulty arises with repeating fractions, but otherwise the method is to shift the fraction to an integer, convert it as above, and then divide by the appropriate power of two in the decimal base. For example:

$$x = 1100.1\overline{01110}\ldots$$
$$x \times 2^6 = 1100101110.\overline{01110}\ldots$$
$$x \times 2 = 11001.\overline{01110}\ldots$$
$$x \times (2^6 - 2) = 1100010101$$
$$x = 1100010101 / 111110$$
$$x = (789 / 62)_{10}$$

Another way of converting from binary to decimal, often quicker for a person familiar with hexadecimal, is to do so indirectly—first converting (x in binary) into (x in hexadecimal) and then converting (x in hexadecimal) into (x in decimal).

For very large numbers, these simple methods are inefficient because they perform a large number of multiplications or divisions where one operand is very large. A simple divide-and-conquer algorithm is more effective asymptotically: given a binary number, it is divided by 10^k, where k is chosen so that the quotient roughly equals the remainder; then each of these pieces is converted to decimal and the two are concatenated. Given a decimal number, it can be split into two pieces of about the same size, each of which is converted to binary, whereupon the first converted piece is multiplied by 10^k and added to the second converted piece, where k is the number of decimal digits in the second, least-significant piece before conversion.

Hexadecimal

0_{hex}	=	0_{dec}	=	0_{oct}	0	0	0	0
1_{hex}	=	1_{dec}	=	1_{oct}	0	0	0	1
2_{hex}	=	2_{dec}	=	2_{oct}	0	0	1	0
3_{hex}	=	3_{dec}	=	3_{oct}	0	0	1	1
4_{hex}	=	4_{dec}	=	4_{oct}	0	1	0	0
5_{hex}	=	5_{dec}	=	5_{oct}	0	1	0	1
6_{hex}	=	6_{dec}	=	6_{oct}	0	1	1	0
7_{hex}	=	7_{dec}	=	7_{oct}	0	1	1	1
8_{hex}	=	8_{dec}	=	10_{oct}	1	0	0	0
9_{hex}	=	9_{dec}	=	11_{oct}	1	0	0	1
A_{hex}	=	10_{dec}	=	12_{oct}	1	0	1	0
B_{hex}	=	11_{dec}	=	13_{oct}	1	0	1	1
C_{hex}	=	12_{dec}	=	14_{oct}	1	1	0	0
D_{hex}	=	13_{dec}	=	15_{oct}	1	1	0	1
E_{hex}	=	14_{dec}	=	16_{oct}	1	1	1	0
F_{hex}	=	15_{dec}	=	17_{oct}	1	1	1	1

Binary may be converted to and from hexadecimal more easily. This is because the radix of the hexadecimal system (16) is a power of the radix of the binary system (2). More specifically, $16 = 2^4$, so it takes four digits of binary to represent one digit of hexadecimal, as shown in the adjacent table.

To convert a hexadecimal number into its binary equivalent, simply substitute the corresponding binary digits:

$3A_{16} = 0011\ 1010_2$

$E7_{16} = 1110\ 0111_2$

To convert a binary number into its hexadecimal equivalent, divide it into groups of four bits. If the number of bits isn't a multiple of four, simply insert extra 0 bits at the left (called padding). For example:

$1010010_2 = 0101\ 0010$ grouped with padding $= 52_{16}$

$11011101_2 = 1101\ 1101$ grouped $= DD_{16}$

To convert a hexadecimal number into its decimal equivalent, multiply the decimal equivalent of each hexadecimal digit by the corresponding power of 16 and add the resulting values:

$C0E7_{16} = (12 \times 16^3) + (0 \times 16^2) + (14 \times 16^1) + (7 \times 16^0) = (12 \times 4096) + (0 \times 256) + (14 \times 16) + (7 \times 1) = 49{,}383_{10}$

Octal

Binary is also easily converted to the octal numeral system, since octal uses a radix of 8, which is a power of two (namely, 2^3, so it takes exactly three binary digits to represent an octal digit). The correspondence between octal and binary numerals is the same as for the first eight digits of hexadecimal in the table above. Binary 000 is equivalent to the octal digit 0, binary 111 is equivalent to octal 7, and so forth.

Octal	Binary
0	000
1	001
2	010
3	011
4	100
5	101
6	110
7	111

Converting from octal to binary proceeds in the same fashion as it does for hexadecimal:

$65_8 = 110\ 101_2$

$17_8 = 001\ 111_2$

And from binary to octal:

$101100_2 = 101\ 100_2$ grouped $= 54_8$

$10011_2 = 010\ 011_2$ grouped with padding $= 23_8$

And from octal to decimal:

$$65_8 = (6 \times 8^1) + (5 \times 8^0) = (6 \times 8) + (5 \times 1) = 53_{10}$$

$$127_8 = (1 \times 8^2) + (2 \times 8^1) + (7 \times 8^0) = (1 \times 64) + (2 \times 8) + (7 \times 1) = 87_{10}$$

Representing Real Numbers

Non-integers can be represented by using negative powers, which are set off from the other digits by means of a radix point (called a decimal point in the decimal system). For example, the binary number 11.01_2 thus means:

1×2^1	$(1 \times 2 = 2)$	plus
1×2^0	$(1 \times 1 = 1)$	plus
0×2^{-1}	$(0 \times \frac{1}{2} = 0)$	plus
1×2^{-2}	$(1 \times \frac{1}{4} = 0.25)$	

For a total of 3.25 decimal.

All dyadic rational numbers $\frac{p}{2^a}$ have a *terminating* binary numeral—the binary representation has a finite number of terms after the radix point. Other rational numbers have binary representation, but instead of terminating, they *recur*, with a finite sequence of digits repeating indefinitely. For instance:

$$\frac{1_{10}}{3_{10}} = \frac{1_2}{11_2} = 0.0101010101\ldots_2$$

$$\frac{12_{10}}{17_{10}} = \frac{1100_2}{10001_2} = 0.10110100\ 10110100\ 10110100\ldots_2$$

The phenomenon that the binary representation of any rational is either terminating or recurring also occurs in other radix-based numeral systems. for instance, the explanation in decimal. Another similarity is the existence of alternative representations for any terminating representation, relying on the fact that 0.111111... is the sum of the geometric series $2^{-1} + 2^{-2} + 2^{-3} + \ldots$ which is 1.

Binary numerals which neither terminate nor recur represent irrational numbers. For instance,

- 0.10100100010000100000100... does have a pattern, but it is not a fixed-length recurring pattern, so the number is irrational.
- 1.0110101000001001111001100110011111110... is the binary representation of $\sqrt{2}$, the square root of 2, another irrational. It has no discernible pattern.

Binary Code

Binary code is the simplest form of computer code or programming data. It is represented entirely by a binary system of digits consisting of a string of consecutive zeros and ones. Binary code is often associated with machine code in that binary sets can be combined to form raw code, which is interpreted by a computer or other piece of hardware.

Importance of Binary Code

In computers and other computerized devices (like calculators, printers, coffee makers, and microwaves), bits are usually transmitted electronically. But this electronic information is fleeting. For it to exist for any length of time—and without a power supply—it must be stored physically within the device's hardware. That means that every piece of binary code in a computer must be converted into a physical object or state. Binary code, as it turns out, is easy to convert from electronic information (e.g., 0s and 1s) to physical information, because only two kinds of physical objects or states are needed.

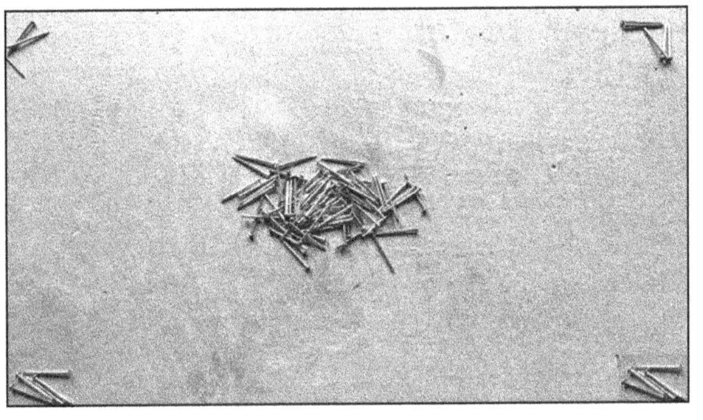

The conversion from electrical information to physical information storage is analogous to someone speaking the binary code for "dog" in 0s and 1s while you write them down on a piece of paper. The spoken 0s and 1s can't be heard indefinitely after they are spoken, but by writing them physically on a piece of paper, you can refer back to them over and over again. In the case of a computer, that binary code might be stored in high and low voltages, in magnetized or demagnetized segments of a metallic disk, or, in super old computers, in punched and unpunched holes in cardstock.

Regardless of the medium, binary code has been the gold standard of physical information storage in computational devices from calculators to super computers.

Importance of Bit Number

Arranging and reading bits in ordered groups is what makes binary exceptionally powerful for storing and transmitting huge amounts of information. To understand why, it helps to consider the alternative: what if only one bit was used at a time? Well, you'd only be able to share two types of information—one type represented by 0 and the other by 1. Forget encoding the entire alphabet or punctuation signs—you just get two kinds of information.

But when you group bits by two, you get four kinds of information:

00, 01, 10, 11

By increasing from two-bit groups to three-bit groups, you double the amount of information you can encode:

000, 001, 010, 011, 100, 101, 110, 111

While eight different kinds of information are still not enough for representing a whole alphabet, perhaps you can see where the pattern is headed.

Using any binary code representation you'd like, try to figure out how many possible combinations of bits you can make out using bits grouped by four. Then try again using bits grouped by five. How many possible combinations do you think you can get using six bits at a time, or 64? By grouping single bits together in larger and larger groups, computers can use binary code to find, organize, send, and store more and more kinds of information.

As computer technology has advanced, computer engineers have needed ways of sending and storing greater amounts of information at a time. As a result, the bit-length used by computers has been growing steadily over the course of computer history. If you have a new iPhone, it is using a 64-bit microprocessor, which means that it stores and accesses information in groups of 64 binary digits—which means that it's capable of storing 2^{64}, or more than 18,000,000,000,000,000,000 unique 64-bit combinations of binary integers.

This idea of coding information with more bits at a time to improve the power and efficiency of computers has driven computer engineering from the beginning, and still does. Though this excerpt from The Soul of a New Machine was first published in 1981, the basic principle of encoding information in binary code with increasing complexity is still representative of the progression of computational power today:

> "Inside certain crucial parts of a typical modern computer, the bits – the electrical symbols – are handled in packets. Like phone numbers the packets are of a standard size. IBM's machines have traditionally handled information in packages 32 bits long. Data General's NOVA and most minicomputers after it, including the Eclipses, deal with packages only 16 bits long. The distinction is inconsequential in theory, since any computer is hypothetically capable of doing what any other computer may do. But the ease and speed with which different computers can be made to perform the same piece of work vary widely, and in general a machine that handles symbols in chunks of 32 bits runs faster, and for some purposes – usually large ones – it is easier to program than a machine that handles only 16 bits at a time."

Advantages of Binary Code

Following is the list of advantages that binary code offers.

- Binary codes are suitable for the computer applications.
- Binary codes are suitable for the digital communications.
- Binary codes make the analysis and designing of digital circuits if we use the binary codes.
- Since only 0 & 1 are being used, implementation becomes easy.

Weighted Codes

Weighted binary codes are those binary codes which obey the positional weight principle. Each position of the number represents a specific weight. Several systems of the codes are used to express the decimal digits 0 through 9. In these codes each decimal digit is represented by a group of four bits.

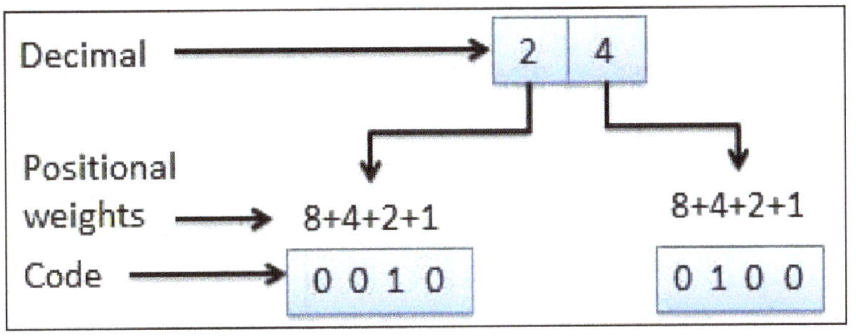

Non-Weighted Codes

In this type of binary codes, the positional weights are not assigned. The examples of non-weighted codes are Excess-3 code and Gray code.

Excess-3 code

The Excess-3 code is also called as XS-3 code. It is non-weighted code used to express decimal numbers. The Excess-3 code words are derived from the 8421 BCD code words adding (0011)2 or (3)10 to each code word in 8421. The excess-3 codes are obtained as follows –

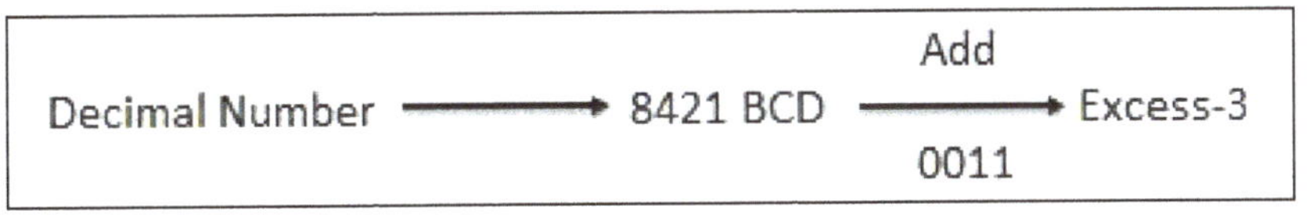

Example;

Decimal	BCD 8 4 2 1	Excess-3 BCD + 0011
0	0 0 0 0	0 0 1 1
1	0 0 0 1	0 1 0 0
2	0 0 1 0	0 1 0 1
3	0 0 1 1	0 1 1 0
4	0 1 0 0	0 1 1 1
5	0 1 0 1	1 0 0 0
6	0 1 1 0	1 0 0 1
7	0 1 1 1	1 0 1 0
8	1 0 0 0	1 0 1 1
9	1 0 0 1	1 1 0 0

Gray Code

It is the non-weighted code and it is not arithmetic codes. That means there are no specific weights assigned to the bit position. It has a very special feature that, only one bit will change each time the decimal number is incremented as shown in fig. As only one bit changes at a time, the gray code is called as a unit distance code. The gray code is a cyclic code. Gray code cannot be used for arithmetic operation.

Decimal	BCD	Gray
0	0 0 0 0	0 0 0 0
1	0 0 0 1	0 0 0 1
2	0 0 1 0	0 0 1 1
3	0 0 1 1	0 0 1 0
4	0 1 0 0	0 1 1 0
5	0 1 0 1	0 1 1 1
6	0 1 1 0	0 1 0 1
7	0 1 1 1	0 1 0 0
8	1 0 0 0	1 1 0 0
9	1 0 0 1	1 1 0 1

Application of Gray Code

- Gray code is popularly used in the shaft position encoders.
- A shaft position encoder produces a code word which represents the angular position of the shaft.

Alphanumeric Codes

A binary digit or bit can represent only two symbols as it has only two states '0' or '1'. But this is not enough for communication between two computers because there we need many more symbols for communication. These symbols are required to represent 26 alphabets with capital and small letters, numbers from 0 to 9, punctuation marks and other symbols.

The alphanumeric codes are the codes that represent numbers and alphabetic characters. Mostly such codes also represent other characters such as symbol and various instructions necessary for conveying information. An alphanumeric code should at least represent 10 digits and 26 letters of alphabet i.e. total 36 items. The following three alphanumeric codes are very commonly used for the data representation.

- American Standard Code for Information Interchange (ASCII).
- Extended Binary Coded Decimal Interchange Code (EBCDIC).
- Five bit Baudot Code.

ASCII code is a 7-bit code whereas EBCDIC is an 8-bit code. ASCII code is more commonly used worldwide while EBCDIC is used primarily in large IBM computers.

Binary-Coded Decimal

Binary Coded Decimal or BCD as it is more commonly called, is another process for converting decimal numbers into their binary equivalents.

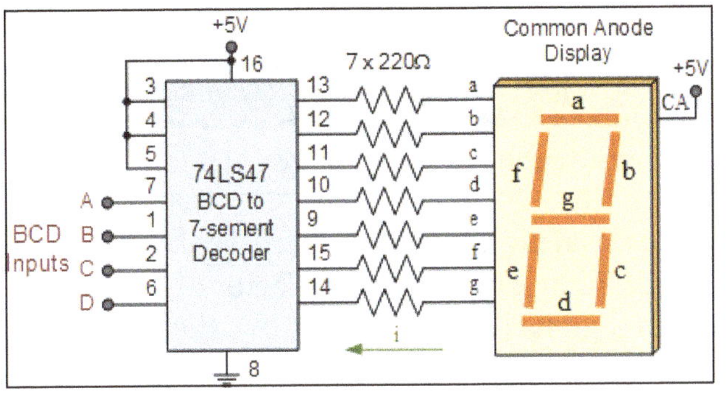

As we have seen in this Binary Numbers section of topic, there are many different binary codes used in digital and electronic circuits, each with its own specific use.

As we naturally live in a decimal (base-10) world we need some way of converting these decimal numbers into a binary (base-2) environment that computers and digital electronic devices understand, and binary coded decimal code allows us to do that.

We have seen previously that an n-bit binary code is a group of "n" bits that assume up to 2^n distinct combinations of 1's and 0's. The advantage of the Binary Coded Decimal system is that each decimal digit is represented by a group of 4 binary digits or bits in much the same way as Hexadecimal. So for the 10 decimal digits (0-to-9) we need a 4-bit binary code.

But do not get confused, binary coded decimal is not the same as hexadecimal. Whereas a 4-bit hexadecimal number is valid up to F_{16} representing binary 1111_2, (decimal 15), binary coded decimal numbers stop at 9 binary 1001_2. This means that although 16 numbers (2^4) can be represented using four binary digits, in the BCD numbering system the six binary code combinations of: 1010 (decimal 10), 1011 (decimal 11), 1100 (decimal 12), 1101 (decimal 13), 1110 (decimal 14), and 1111 (decimal 15) are classed as forbidden numbers and can not be used.

The main advantage of binary coded decimal is that it allows easy conversion between decimal (base-10) and binary (base-2) form. However, the disadvantage is that BCD code is wasteful as the states between 1010(decimal 10), and 1111 (decimal 15) are not used. Nevertheless, binary coded decimal has many important applications especially using digital displays.

In the BCD numbering system, a decimal number is separated into four bits for each decimal digit within the number. Each decimal digit is represented by its weighted binary value performing a direct translation of the number. So a 4-bit group represents each displayed decimal digit from 0000 for a zero to 1001 for a nine.

So for example, 357_{10} (Three Hundred and Fifty Seven) in decimal would be presented in Binary Coded Decimal as:

$$357_{10} = 0011\ 0101\ 0111\ (BCD)$$

Then we can see that BCD uses weighted codification, because the binary bit of each 4-bit group represents a given weight of the final value. In other words, the BCD is a weighted code and the weights used in binary coded decimal code are 8, 4, 2, 1, commonly called the 8421 code as it forms the 4-bit binary representation of the relevant decimal digit.

Binary Coded Decimal Representation of a Decimal Number

Binary Power	2^3	2^2	2^1	2^0
Binary Weight:	8	4	2	1

The decimal weight of each decimal digit to the left increases by a factor of 10. In the BCD number system, the binary weight of each digit increases by a factor of 2 as shown. Then the first digit has a weight of 1 (2^0), the second digit has a weight of 2 (2^1), the third a weight of 4 (2^2), the fourth a weight of 8 (2^3).

Then the relationship between decimal (denary) numbers and weighted binary coded decimal digits is given below.

Truth Table for Binary Coded Decimal

Decimal Number	BCD 8421 Code
0	0000 0000
1	0000 0001
2	0000 0010
3	0000 0011
4	0000 0100
5	0000 0101
6	0000 0110
7	0000 0111
8	0000 1000
9	0000 1001
10 (1+0)	0001 0000
11 (1+1)	0001 0001
12 (1+2)	0001 0010
...	...
20 (2+0)	0010 0000
21 (2+1)	0010 0001
22 (2+2)	0010 0010
etc., continuing upwards in groups of four	

Then we can see that 8421 BCD code is nothing more than the weights of each binary digit, with each decimal (denary) number expressed as its four-bit pure binary equivalent.

Decimal-to-BCD Conversion

As we have seen above, the conversion of decimal to binary coded decimal is very similar to the conversion of hexadecimal to binary. Firstly, separate the decimal number into its weighted digits and then write down the equivalent 4-bit 8421 BCD code representing each decimal digit as shown.

Binary Coded Decimal Example

Using the above table, convert the following decimal (denary) numbers: 85_{10}, 572_{10} and 8579_{10} into their 8421 BCD equivalents.

8510 = 1000 0101 (BCD)

57210 = 0101 0111 0010 (BCD)

857910 = 1000 0101 0111 1001 (BCD)

Note that the resulting binary number after the conversion will be a true binary translation of decimal digits. This is because the binary code translates as a true binary count.

BCD-to-Decimal Conversion

The conversion from binary coded decimal to decimal is the exact opposite of the above. Simply divide the binary number into groups of four digits, starting with the least significant digit and then write the decimal digit represented by each 4-bit group. Add additional zero's at the end if required to produce a complete 4-bit grouping. So for example, 110101_2 would become: $0011\ 0101_2$ or 35_{10} in decimal.

Binary Coded Decimal Example

Convert the following binary numbers: 1001_2, 1010_2, 1000111_2 and 10100111000.101_2 into their decimal equivalents.

$1001_2 = 1001_{BCD} = 9_{10}$

1010_2 = this will produce an error as it is decimal 10_{10} and not a valid BCD number

$1000111_2 = 0100\ 0111_{BCD} = 47_{10}$

$10100111000.101_2 = 0101\ 0011\ 0001.1010_{BCD} = 538.625_{10}$

The conversion of BCD-to-decimal or decimal-to-BCD is a relatively straight forward task but we need to remember that BCD numbers are decimal numbers and not binary numbers, even though they are represented using bits. The BCD representation of a decimal number is important to understand, because microprocessor based systems used by most people needs to be in the decimal system.

However, while BCD is easy to code and decode, it is not an efficient way to store numbers. In the standard 8421 BCD encoding of decimal numbers, the number of individual data bits needed to represent a given decimal number will always be greater than the number of bits required for an equivalent binary encoding.

For example, in binary a three digit decimal number from 0-to-999 requires only 10-bits (1111100111_2), whereas in binary coded decimal, the same number requires a minimum of 12-bits ($0011\ 1110\ 0111_{BCD}$) for the same representation.

Also, performing arithmetic tasks using binary coded decimal numbers can be a bit awkward since each digit can not exceed 9. The addition of two decimal digits in BCD, will create a possible carry bit of 1 which needs to be added to the next group of 4-bits.

If the binary sum with the added carry bit is equal to or less than 9 (1001), the corresponding BCD digit is correct. But when the binary sum is greater than 9 the result is an invalid BCD digit. Therefore it is better to convert BCD numbers into pure binary, perform the required addition, and then convert the back to BCD before displaying the results.

Nevertheless, the use of a BCD coding system in both microelectronics and computer systems is particularly useful in situations where the binary coded decimal is intended to be displayed on one or more 7-segment LED or LCD displays and there are many popular integrated circuits available that are configured to give a BCD output or outputs.

One common IC is the 74LS90 asynchronous counter/divider that contains independent divide-by-2 and divide-by-5 counters that can be used together to produce a divide-by-10 decade counter with BCD outputs. Another is the 74LS390 which is a dual version of the basic 74LS90, and can also be configured to produce a BCD output.

But the most commonly used BCD encoded IC's are the 74LS47 and the 74LS48 BCD to 7-segment decoder/driver, which converts a 4-bit BCD code of a counter, etc. and converts it into the required display code to drive the individual segments of a 7-segment LED display. While both IC's are functionally the same, the 74LS47 has active-low outputs for driving common-anode displays, while the 74LS48 has active-high outputs for driving common-cathode displays.

Binary Coded Decimal Decoder IC

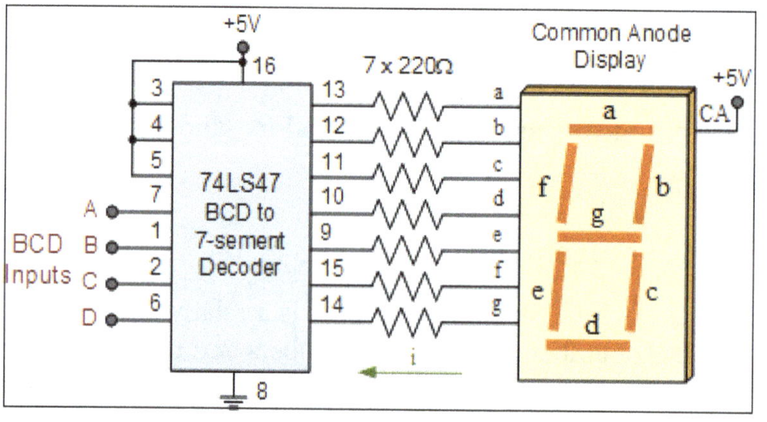

Binary Coded Decimal

We have seen here that Binary Coded Decimal or BCD is simply the 4-bit binary code representation of a decimal digit with each decimal digit replaced in the integer and fractional parts with its binary equivalent. BCD Code uses four bits to represent the 10 decimal digits of 0 to 9.

So for example, if we wanted to display decimal numbers in the range of 0-to-9, (one digit) we would need 4 data bits (a nibble), decimal numbers in the range of 0-to-99, (two digits) we would need 8 bits (one byte), decimal numbers in the range of 0-to-999, (three digits) we would need 12 bits, and so on. The use of a single byte (8-bits) to store or display two BCD digits, allowing a byte to hold a BCD number in the range of 00 – 99, is known as packed BCD.

Standard binary coded decimal code is commonly known as a weighted 8421 BCD code, with 8, 4, 2 and 1 representing the weights of the different bits starting from the most significant bit (MSB) and proceeding towards the least significant bit (LSB). The weights of the individual positions of the bits of a BCD code are: $2^3 = 8$, $2^2 = 4$, $2^1 = 2$, $2^0 = 1$.

The main advantage of the Binary Coded Decimal system is that it is a fast and efficient system to convert the decimal numbers into binary numbers as compared to the pure binary system. But the BCD code is wasteful as many of the 4-bit states (10-to-16) are not used but decimal displays have important applications.

Bitwise Operation

A bitwise operator is an operator used to perform bitwise operations on bit patterns or binary numerals that involve the manipulation of individual bits.

Bitwise operators are used in:

- Communication stacks where the individual bits in the header attached to the data signify important information
- Embedded software for controlling different functions in the chip and indicating the status of hardware by manipulating the individual bits of hardware registers of embedded microcontrollers
- Low-level programming for applications such as device drivers, cryptographic software, video decoding software, memory allocators, compression software and graphics
- Maintaining large sets of integers efficiently in search and optimization problems
- Bitwise operations performed on bit flags, which can enable an instance of enumeration type to store any combination of values defined in an enumerator list

Bitwise operators are used to change individual bits in an operand. A single byte of computer memory-when viewed as 8 bits-can signify the true/false status of 8 flags because each bit can be used as a boolean variable that can hold one of two values: true or false. A *flag* variable is typically used to indicate the status of something. For instance, computer files can be marked as read-only. So you might have a $fReadOnly variable whose job would be to hold the read-only status of a file. This variable is called a flag variable because when $fReadOnly has a true value, it's equivalent to a football referee throwing a flag. The variable says, "Whoa! Don't modify this file."

When you have more than one flag variable, it might be more efficient to use a single variable to indicate the value of more than one flag.

Bitwise Operators

In the explanations below, any indication of a bit's position is counted from the right (least significant) side, advancing left. For example, the binary value 0001 (decimal 1) has zeroes at every position but the first one.

NOT

The bitwise NOT, or complement, is a unary operation that performs logical negation on each bit, forming the ones' complement of the given binary value. Bits that are 0 become 1, and those that are 1 become 0. For example:

NOT 0111 (decimal 7)

= 1000 (decimal 8)

NOT 10101011

= 01010100

The bitwise complement is equal to the two's complement of the value minus one. If two's complement arithmetic is used, then NOT x = -x − 1.

For unsigned integers, the bitwise complement of a number is the "mirror reflection" of the number across the half-way point of the unsigned integer's range. For example, for 8-bit unsigned integers, NOT x = 255 - x, which can be visualized on a graph as a downward line that effectively "flips" an increasing range from 0 to 255, to a decreasing range from 255 to 0. A simple but illustrative example use is to invert a grayscale image where each pixel is stored as an unsigned integer.

AND

A bitwise AND takes two equal-length binary representations and performs the logical AND operation on each pair of the corresponding bits, by multiplying them. Thus, if both bits in the compared position are 1, the bit in the resulting binary representation is 1 (1 × 1 = 1); otherwise, the result is 0 (1 × 0 = 0 and 0 × 0 = 0). For example:

0101 (decimal 5)

AND 0011 (decimal 3)

= 0001 (decimal 1)

The operation may be used to determine whether a particular bit is *set* (1) or *clear* (0). For example, given a bit pattern 0011 (decimal 3), to determine whether the second bit is set we use a bitwise AND with a bit pattern containing 1 only in the second bit:

0011 (decimal 3)

AND 0010 (decimal 2)

= 0010 (decimal 2)

Because the result 0010 is non-zero, we know the second bit in the original pattern was set. This is often called *bit masking*. (By analogy, the use of masking tape covers, or *masks*, portions that should not be altered or portions that are not of interest. In this case, the 0 values mask the bits that are not of interest).

The bitwise AND may be used to clear selected bits (or flags) of a register in which each bit represents an individual Boolean state. This technique is an efficient way to store a number of Boolean values using as little memory as possible.

For example, 0110 (decimal 6) can be considered a set of four flags, where the first and fourth flags are clear (0), and the second and third flags are set (1). The second bit may be cleared by using a bitwise AND with the pattern that has a zero only in the second bit:

0110 (decimal 6)

AND 1101 (decimal 13)

= 0100 (decimal 4)

Because of this property, it becomes easy to check the parity of a binary number by checking the value of the lowest valued bit. Using the example above:

0110 (decimal 6)

AND 0001 (decimal 1)

= 0000 (decimal 0)

Because 6 AND 1 is zero, 6 is divisible by two and therefore even.

OR

A bitwise OR takes two bit patterns of equal length and performs the logical inclusive OR operation on each pair of corresponding bits. The result in each position is 0 if both bits are 0, while otherwise the result is 1. For example:

0101 (decimal 5)

OR 0011 (decimal 3)

= 0111 (decimal 7)

The bitwise OR may be used to set to 1 the selected bits of the register described above. For example, the fourth bit of 0010 (decimal 2) may be set by performing a bitwise OR with the pattern with only the fourth bit set:

0010 (decimal 2)

OR 1000 (decimal 8)

= 1010 (decimal 10)

XOR

A bitwise XOR takes two bit patterns of equal length and performs the logical exclusive OR operation on each pair of corresponding bits. The result in each position is 1 if only the first bit is 1 *or* only the second bit is 1, but will be 0 if both are 0 or both are 1. In this we perform the comparison of two bits, being 1 if the two bits are different, and 0 if they are the same. For example:

0101 (decimal 5)

XOR 0011 (decimal 3)

= 0110 (decimal 6)

The bitwise XOR may be used to invert selected bits in a register (also called toggle or flip). Any bit may be toggled by XORing it with 1. For example, given the bit pattern 0010 (decimal 2) the second and fourth bits may be toggled by a bitwise XOR with a bit pattern containing 1 in the second and fourth positions:

0010 (decimal 2)

XOR 1010 (decimal 10)

= 1000 (decimal 8)

This technique may be used to manipulate bit patterns representing sets of Boolean states.

Assembly language programmers and optimizing compilers sometimes use XOR as a short-cut to setting the value of a register to zero. Performing XOR on a value against itself always yields zero, and on many architectures this operation requires fewer clock cycles and memory than loading a zero value and saving it to the register.

Mathematical Equivalents

Assuming $x \geq y$, for the non-negative integers, the bitwise operations can be written as follows:

$$\text{NOT}\, x = \sum_{n=0}^{\lfloor \log_2(x) \rfloor} 2^n \left[\left(\left\lfloor \frac{x}{2^n} \right\rfloor \bmod 2 + 1 \right) \bmod 2 \right] = 2^{\lfloor \log_2(x) \rfloor + 1} - 1 - x$$

$$x\, \text{AND}\, y = \sum_{n=0}^{\lfloor \log_2(x) \rfloor} 2^n \left(\left\lfloor \frac{x}{2^n} \right\rfloor \bmod 2 \right) \left(\left\lfloor \frac{y}{2^n} \right\rfloor \bmod 2 \right)$$

$$x\, \text{OR}\, y = \sum_{n=0}^{\lfloor \log_2(x) \rfloor} 2^n \left[\left[\left(\left\lfloor \frac{x}{2^n} \right\rfloor \bmod 2 \right) + \left(\left\lfloor \frac{y}{2^n} \right\rfloor \bmod 2 \right) + \left(\left\lfloor \frac{x}{2^n} \right\rfloor \bmod 2 \right) \left(\left\lfloor \frac{y}{2^n} \right\rfloor \bmod 2 \right) \right] \bmod 2 \right]$$

$$x\, \text{XOR}\, y = \sum_{n=0}^{\lfloor \log_2(x) \rfloor} 2^n \left[\left[\left(\left\lfloor \frac{x}{2^n} \right\rfloor \bmod 2 \right) + \left(\left\lfloor \frac{y}{2^n} \right\rfloor \bmod 2 \right) \right] \bmod 2 \right] \sum_{n=0}^{\lfloor \log_2(x) \rfloor} 2^n \left[\left(\left\lfloor \frac{x}{2^n} \right\rfloor + \left\lfloor \frac{y}{2^n} \right\rfloor \right) \bmod 2 \right]$$

Bit Shifts

The bit shifts are sometimes considered bitwise operations, because they treat a value as a series of bits rather than as a numerical quantity. In these operations the digits are moved, or *shifted*, to the left or right. Registers in a computer processor have a fixed width, so some bits will be "shifted out" of the register at one end, while the same number of bits are "shifted in" from the other end; the differences between bit shift operators lie in how they determine the values of the shifted-in bits.

Arithmetic Shift

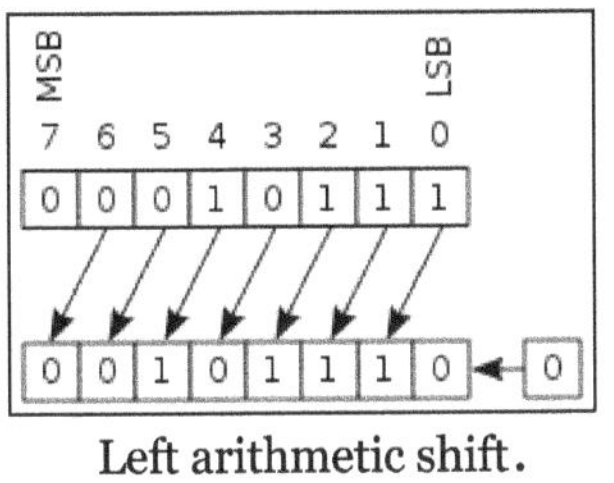

Left arithmetic shift.

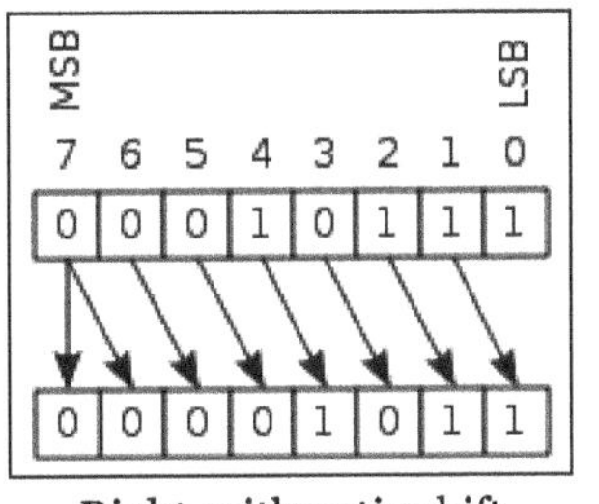

Right arithmetic shift.

In an *arithmetic shift*, the bits that are shifted out of either end are discarded. In a left arithmetic shift, zeros are shifted in on the right; in a right arithmetic shift, the sign bit (the MSB in two's complement) is shifted in on the left, thus preserving the sign of the operand.

This example uses an 8-bit register:

00010111 (decimal +23) LEFT-SHIFT

= 00101110 (decimal +46)

10010111 (decimal −105) RIGHT-SHIFT

= 11001011 (decimal −53)

In the first case, the leftmost digit was shifted past the end of the register, and a new 0 was shifted into the rightmost position. In the second case, the rightmost 1 was shifted out (perhaps into the carry flag), and a new 1 was copied into the leftmost position, preserving the sign of the number. Multiple shifts are sometimes shortened to a single shift by some number of digits. For example:

00010111 (decimal +23) LEFT-SHIFT-BY-TWO

= 01011100 (decimal +92)

A left arithmetic shift by n is equivalent to multiplying by 2^n (provided the value does not overflow), while a right arithmetic shift by n of a two's complement value is equivalent to dividing by 2^n and rounding toward negative infinity. If the binary number is treated as ones' complement, then the same right-shift operation results in division by 2^n and rounding toward zero.

Logical Shift

In a *logical shift*, zeros are shifted in to replace the discarded bits. Therefore, the logical and arithmetic left-shifts are exactly the same.

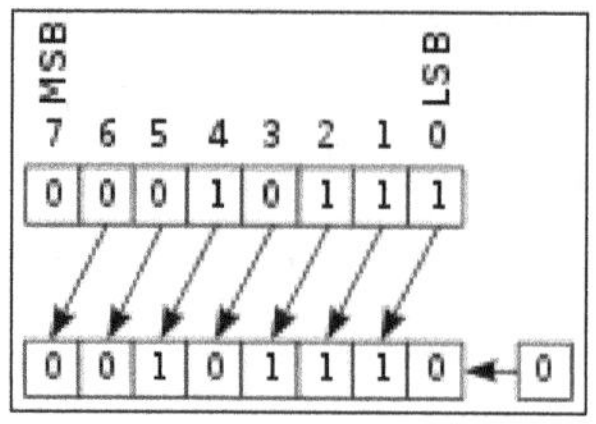

Left logical shift.

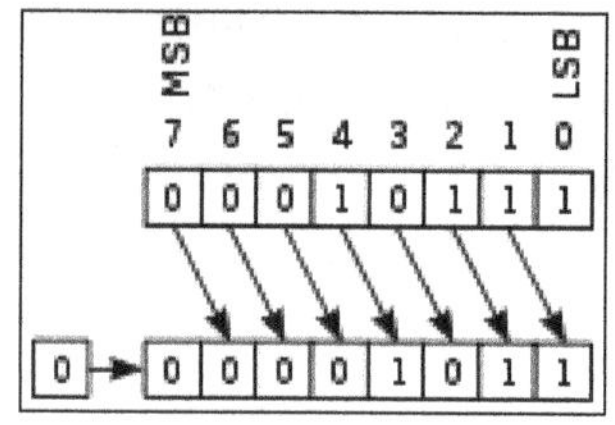

Right logical shift.

However, as the logical right-shift inserts value 0 bits into the most significant bit, instead of copying the sign bit, it is ideal for unsigned binary numbers, while the arithmetic right-shift is ideal for signed two's complement binary numbers.

Circular Shift

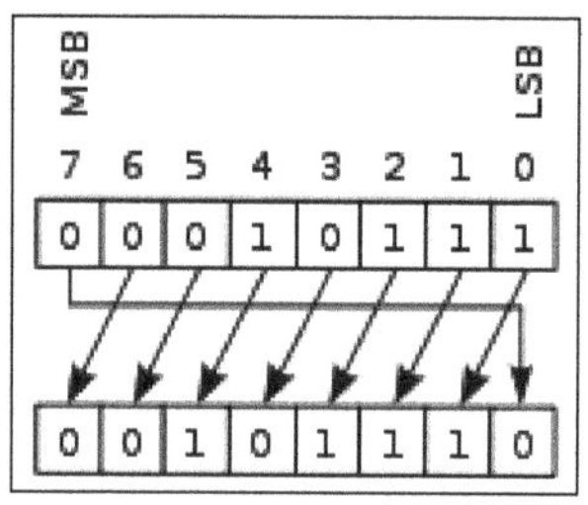

Left circular shift or rotate.

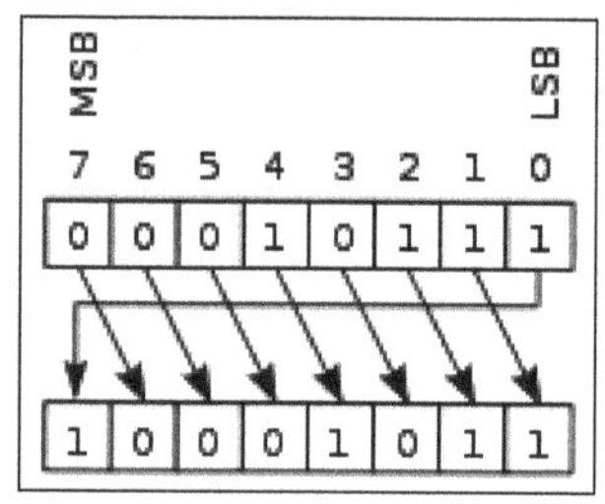

Right circular shift or rotate.

Another form of shift is the *circular shift* or *bit rotation*. In this operation, the bits are "rotated" as if the left and right ends of the register were joined. The value that is shifted in on the right during a left-shift is whatever value was shifted out on the left, and vice versa. This operation is useful if it is necessary to retain all the existing bits, and is frequently used in digital cryptography.

Rotate through Carry

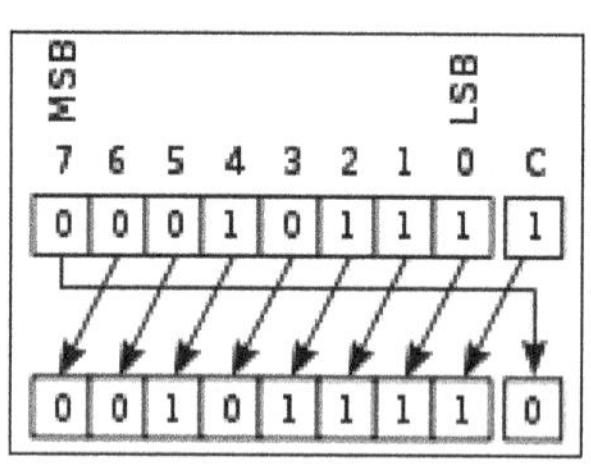

Left rotate through carry.

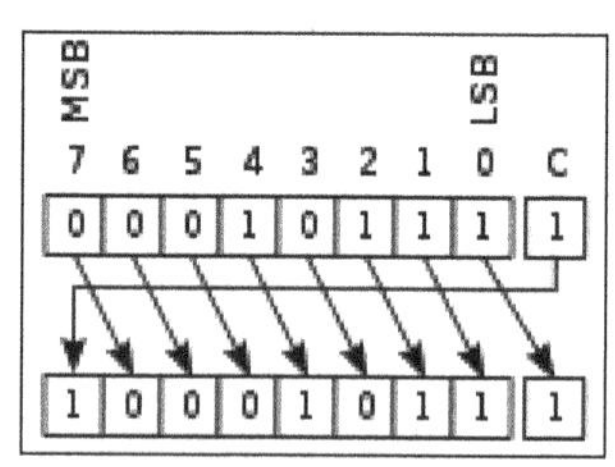

Right rotate through carry.

Rotate through carry is similar to the *rotate no carry* operation, but the two ends of the register are separated by the carry flag. The bit that is shifted in (on either end) is the old value of the carry flag, and the bit that is shifted out (on the other end) becomes the new value of the carry flag.

A single *rotate through carry* can simulate a logical or arithmetic shift of one position by setting up the carry flag beforehand. For example, if the carry flag contains 0, then x RIGHT-ROTATE-THROUGH-CARRY-BY-ONE is a logical right-shift, and if the carry flag contains a copy of the sign bit, then x RIGHT-ROTATE-THROUGH-CARRY-BY-ONE is an arithmetic right-shift. For this reason, some microcontrollers such as low end PICs just have *rotate* and *rotate through carry*, and don't bother with arithmetic or logical shift instructions.

Rotate through carry is especially useful when performing shifts on numbers larger than the processor's native word size, because if a large number is stored in two registers, the bit that is shifted off the end of the first register must come in at the other end of the second. With rotate-through-carry, that bit is "saved" in the carry flag during the first shift, ready to shift in during the second shift without any extra preparation.

Shifts in C, C++, C#, Go, Java, JavaScript, Pascal, Perl, PHP, Python and Ruby

In C-inspired languages, the left and right shift operators are "<<" and ">>", respectively. The number of places to shift is given as the second argument to the shift operators. For example,

```
x = y << 2;
```

assigns x the result of shifting y to the left by two bits, which is equivalent to a multiplication by four.

Shifts can result in implementation-defined behavior or undefined behavior, so care must be taken when using them. The result of shifting by a bit count greater than or equal to the word's size is undefined behavior in C and C++. Right-shifting a negative value is implementation-defined and not recommended by good coding practice; the result of left-shifting a signed value is undefined if the result cannot be represented in the result type. In C#, the right-shift is an arithmetic shift when the first operand is an int or long. If the first operand is of type uint or ulong, the right-shift is a logical shift.

Circular Shifts in C-family Languages

The C-family of languages lack a rotate operator, but one can be synthesized from the shift operators. Care must be taken to ensure the statement is well formed to avoid undefined behavior and timing attacks in software with security requirements. For example, a naive implementation that left rotates a 32-bit unsigned value x by n positions is simply:

```
unsigned int x = ..., n = ...;

unsigned int y = (x << n) | (x >> (32 - n));
```

However, a shift by 0 bits results in undefined behavior in the right hand expression (x >> (32 - n)) because 32 - 0 is 32, and 32 is outside the range [0 - 31] inclusive. A second try might result in:

```
unsigned int x = ..., n = ...;

unsigned int y = n ? (x << n) | (x >> (32 - n)) : x;
```

where the shift amount is tested to ensure it does not introduce undefined behavior. However, the branch adds an additional code path and presents an opportunity for timing analysis and attack, which is often not acceptable in high integrity software. In addition, the code compiles to multiple machine instructions, which is often less efficient than the processor's native instruction.

To avoid the undefined behavior and branches under GCC and Clang, the following should be used. The pattern is recognized by many compilers, and the compiler will emit a single rotate instruction:

```
unsigned int x = ..., n = ...;

unsigned int y = (x << n) | (x >> (-n & 31));
```

There are also compiler-specific intrinsics implementing circular shifts, like _rotl8, _rotl16, _rotr8, _rotr16 in Microsoft Visual C++. Clang provides some rotate intrinsics for Microsoft compatibility that suffers the problems above. GCC does not offer rotate intrinsics. Intel also provides x86 Intrinsics.

Shifts in Java

In Java, all integer types are signed, so the "<<" and ">>" operators perform arithmetic shifts. Java adds the operator ">>>" to perform logical right shifts, but since the logical and arithmetic left-shift operations are identical for signed integer, there is no "<<<" operator in Java.

More details of Java shift operators:

- The operators << (left shift), >> (signed right shift), and >>> (unsigned right shift) are called the *shift operators*.
- The type of the shift expression is the promoted type of the left-hand operand. For example, aByte >>> 2 is equivalent to ((int) aByte) >>> 2.
- If the promoted type of the left-hand operand is int, only the five lowest-order bits of the right-hand operand are used as the shift distance. It is as if the right-hand operand were subjected to a bitwise logical AND operator & with the mask value 0x1f (0b11111). The shift distance actually used is therefore always in the range 0 to 31, inclusive.
- If the promoted type of the left-hand operand is long, then only the six lowest-order bits of the right-hand operand are used as the shift distance. It is as if the right-hand operand were subjected to a bitwise logical AND operator & with the mask value 0x3f (0b111111). The shift distance actually used is therefore always in the range 0 to 63, inclusive.
- The value of n >>> s is *n* right-shifted *s* bit positions with zero-extension.
- In bit and shift operations, the type *byte* is implicitly converted to *int*. If the byte value is negative, the highest bit is one, then ones are used to fill up the extra bytes in the int. So byte b1 = -5; int i = b1 | 0x0200; will give i == -5 as result.

Shifts in JavaScript

JavaScript uses bitwise operations to evaluate each of two or more units place to 1 or 0.

Shifts in Pascal

In Pascal, as well as in all its dialects (such as Object Pascal and Standard Pascal), the left and right shift operators are "shl" and "shr", respectively. The number of places to shift is given as the second argument. For example, the following assigns x the result of shifting y to the left by two bits:

```
x := y shl 2;
```

Applications

Bitwise operations are necessary particularly in lower-level programming such as device drivers, low-level graphics, communications protocol packet assembly, and decoding.

Although machines often have efficient built-in instructions for performing arithmetic and logical operations, all these operations can be performed by combining the bitwise operators and zero-testing in various ways. For example, here is a pseudocode implementation of ancient Egyptian multiplication showing how to multiply two arbitrary integers a and b (a greater than b) using only bitshifts and addition:

```
c ← 0
while b ≠ 0
    if (b and 1) ≠ 0
        c ← c + a
    left shift a by 1
    right shift b by 1
return c
```

Another example is a pseudocode implementation of addition, showing how to calculate a sum of two integers a and b using bitwise operators and zero-testing:

```
while a ≠ 0
    c ← b and a
    b ← b xor a
    left shift c by 1
    a ← c
return b
```

Boolean Algebra

A set of rules or Laws of Boolean Algebra expressions have been invented to help reduce the number of logic gates needed to perform a particular logic operation resulting in a list of functions or theorems known commonly as the Laws of Boolean Algebra.

Boolean Algebra is the mathematics we use to analyse digital gates and circuits. We can use these "Laws of Boolean" to both reduce and simplify a complex Boolean expression in an attempt to reduce the number of logic gates required. Boolean Algebra is therefore a system of mathematics based on logic that has its own set of rules or laws which are used to define and reduce Boolean expressions.

The variables used in Boolean Algebra only have one of two possible values, a logic "0" and a logic "1" but an expression can have an infinite number of variables all labelled individually to represent inputs to the expression, For example, variables A, B, C etc, giving us a logical expression of A + B = C, but each variable can ONLY be a 0 or a 1.

Examples of these individual laws of Boolean, rules and theorems for Boolean Algebra are given in the following table.

Truth Tables for the Laws of Boolean

Boolean Expression	Description	Equivalent Switching Circuit	Boolean Algebra Law or Rule
A + 1 = 1	A in parallel with closed = "CLOSED"		Annulment
A + 0 = A	A in parallel with open = "A"	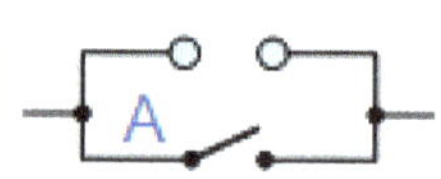	Identity
A . 1 = A	A in series with closed = "A"		Identity
A . 0 = 0	A in series with open = "OPEN"		Annulment
A + A = A	A in parallel with A = "A"	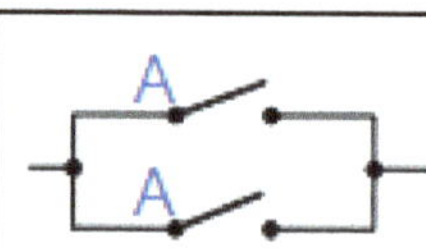	Idempotent
A . A = A	A in series with A = "A"		Idempotent
NOT A = A	NOT NOT A (double negative) = "A"		Double Negation
A + A = 1	A in parallel with NOT A = "CLOSED"	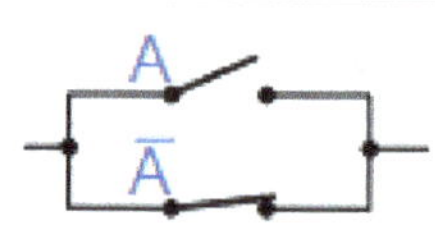	Complement

A . A = 0	A in series with NOT A = "OPEN"		Complement
A+B = B+A	A in parallel with B = B in parallel with A	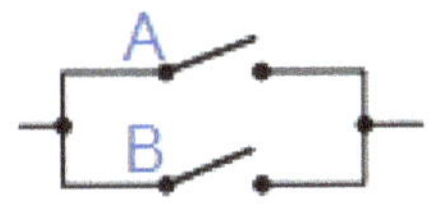	Commutative
A.B = B.A	A in series with B = B in series with A		Commutative
A+B = A.B	invert and replace OR with AND		de Morgan's Theorem
A.B = A+B	invert and replace AND with OR		de Morgan's Theorem

The basic Laws of Boolean Algebra that relate to the Commutative Law allowing a change in position for addition and multiplication, the Associative Law allowing the removal of brackets for addition and multiplication, as well as the Distributive Law allowing the factoring of an expression, are the same as in ordinary algebra.

Each of the Boolean Laws above are given with just a single or two variables, but the number of variables defined by a single law is not limited to this as there can be an infinite number of variables as inputs too the expression. These Boolean laws detailed above can be used to prove any given Boolean expression as well as for simplifying complicated digital circuits.

Operations

Basic Operations

The basic operations of Boolean algebra are as follows:

- AND (conjunction), denoted $x \wedge y$ (sometimes x AND y or Kxy), satisfies $x \wedge y = 1$ if $x = y = 1$ and $x \wedge y = 0$ otherwise.
- OR (disjunction), denoted $x \vee y$ (sometimes x OR y or Axy), satisfies $x \vee y = 0$ if $x = y = 0$ and $x \vee y = 1$ otherwise.
- NOT (negation), denoted $\neg x$ (sometimes NOT x, Nx or $!x$), satisfies $\neg x = 0$ if $x = 1$ and $\neg x = 1$ if $x = 0$.

Alternatively the values of $x \wedge y$, $x \vee y$, and $\neg x$ can be expressed by tabulating their values with truth tables as follows.

x	y	$x \wedge y$	$x \vee y$
0	0	0	0
1	0	0	1
0	1	0	1
1	1	1	1

x	$\neg x$
0	1
1	0

If the truth values 0 and 1 are interpreted as integers, these operations may be expressed with the ordinary operations of arithmetic (where $x + y$ uses addition and xy uses multiplication), or by the minimum/maximum functions:

$$x \wedge y = xy = \min(x, y)$$
$$x \vee y = x + y - xy = \max(x, y)$$
$$\neg x = 1 - x$$

One may consider that only the negation and one of the two other operations are basic, because of the following identities that allow to define the conjunction in terms of the negation and the disjunction, and vice versa:

$$x \wedge y = \neg(\neg x \vee \neg y)$$
$$x \vee y = \neg(\neg x \wedge \neg y)$$

Secondary operations

The three Boolean operations described above are referred to as basic, meaning that they can be taken as a basis for other Boolean operations that can be built up from them by composition, the manner in which operations are combined or compounded. Operations composed from the basic operations include the following examples:

$$x \rightarrow y = \neg x \vee y$$
$$x \oplus y = (x \vee y) \wedge \neg(x \wedge y)$$
$$x \equiv y = \neg(x \oplus y)$$

These definitions give rise to the following truth tables giving the values of these operations for all four possible inputs.

x	y	$x \rightarrow y$	$x \oplus y$	$x \equiv y$
0	0	1	0	1
1	0	0	1	0
0	1	1	1	0
1	1	1	0	1

The first operation, $x \rightarrow y$, or Cxy, is called material implication. If x is true then the value of $x \rightarrow y$ is taken to be that of y. But if x is false then the value of y can be ignored; however the operation must return *some* boolean value and there are only two choices. So by definition, $x \rightarrow y$ is *true* when x is false. (Relevance logic suggests this definition by viewing an implication with a false premise as something other than either true or false.)

The second operation, $x \oplus y$, or Jxy, is called exclusive or (often abbreviated as XOR) to distinguish it from disjunction as the inclusive kind. It excludes the possibility of both x and y. Defined in terms of arithmetic it is addition mod 2 where $1 + 1 = 0$.

The third operation, the complement of exclusive or, is equivalence or Boolean equality: $x \equiv y$, or Exy, is true just when x and y have the same value. Hence $x \oplus y$ as its complement can be understood as $x \neq y$, being true just when x and y are different. Equivalence's counterpart in arithmetic mod 2 is $x + y + 1$.

Given two operands, each with two possible values, there are $2^2 = 4$ possible combinations of inputs. Because each output can have two possible values, there are a total of $2^4 = 16$ possible binary Boolean operations.

Laws

A law of Boolean algebra is an identity such as $x \vee (y \vee z) = (x \vee y) \vee z$ between two Boolean terms, where a Boolean term is defined as an expression built up from variables and the constants 0 and 1 using the operations $\wedge$, $\vee$, and $\neg$. The concept can be extended to terms involving other Boolean operations such as $\oplus$, $\rightarrow$, and $\equiv$, but such extensions are unnecessary for the purposes to which the laws are put. Such purposes include the definition of a Boolean algebra as any model of the Boolean laws, and as a means for deriving new laws from old as in the derivation of $x \vee (y \wedge z) = x \vee (z \wedge y)$ from $y \wedge z = z \wedge y$ as treated in the section on axiomatization.

Monotone Laws

Boolean algebra satisfies many of the same laws as ordinary algebra when one matches up $\vee$ with addition and $\wedge$ with multiplication. In particular the following laws are common to both kinds of algebra:

Associativity of $\vee$:	$x \vee (y \vee z) = (x \vee y) \vee z$
Associativity of $\wedge$:	$x \wedge (y \wedge z) = (x \wedge y) \wedge z$
Commutativity of $\vee$:	$x \vee y = y \vee x$
Commutativity of $\wedge$:	$x \wedge y = y \wedge x$
Distributivity of $\wedge$ over $\vee$:	$x \wedge (y \vee z) = (x \wedge y) \vee (x \wedge z)$
Identity for $\vee$:	$x \vee 0 = x$
Identity for $\wedge$:	$x \wedge 1 = x$
Annihilator for $\wedge$:	$x \wedge 0 = 0$

The following laws hold in Boolean Algebra, but not in ordinary algebra:

Annihilator for $\vee$:	$x \vee 1 = 1$
Idempotence of $\vee$:	$x \vee x = x$
Idempotence of $\wedge$:	$x \wedge x = x$
Absorption 1:	$x \wedge (x \vee y) = x$
Absorption 2:	$x \vee (x \wedge y) = x$
Distributivity of $\vee$ over $\wedge$:	$x \vee (y \wedge z) = (x \vee y) \wedge (x \vee z)$

Taking x = 2 in the third law above shows that it is not an ordinary algebra law, since 2×2 = 4. The remaining five laws can be falsified in ordinary algebra by taking all variables to be 1, for example in Absorption Law 1 the left hand side would be 1(1+1) = 2 while the right hand side would be 1, and so on.

All of the laws treated so far have been for conjunction and disjunction. These operations have the property that changing either argument either leaves the output unchanged or the output changes in the same way as the input. Equivalently, changing any variable from 0 to 1 never results in the output changing from 1 to 0. Operations with this property are said to be monotone. Thus the axioms so far have all been for monotonic Boolean logic. Nonmonotonicity enters via complement ¬ as follows.

Nonmonotone Laws

The complement operation is defined by the following two laws.

Complementation 1 $x \wedge \neg x = 0$

Complementation 2 $x \vee \neg x = 1$

All properties of negation including the laws below follow from the above two laws alone.

In both ordinary and Boolean algebra, negation works by exchanging pairs of elements, whence in both algebras it satisfies the double negation law (also called involution law).

Double negation $\neg(\neg x) = x$

But whereas *ordinary algebra* satisfies the two laws.

$$(-x)(-y) = xy$$
$$(-x) + (-y) = -(x + y)$$

Boolean algebra satisfies De Morgan's laws:

De Morgan1 $\neg x \wedge \neg y = \neg(x \vee y)$

De Morgan 2 $\neg x \vee \neg y = \neg(x \wedge y)$

Completeness

The laws listed above define Boolean algebra, in the sense that they entail the rest of the subject. The laws *Complementation* 1 and 2, together with the monotone laws, suffice for this purpose and can therefore be taken as one possible *complete* set of laws or axiomatization of Boolean algebra. Every law of Boolean algebra follows logically from these axioms. Furthermore, Boolean algebras can then be defined as the models of these axioms as treated in the section thereon.

To clarify, writing down further laws of Boolean algebra cannot give rise to any new consequences of these axioms, nor can it rule out any model of them. In contrast, in a list of some but not all of the same laws, there could have been Boolean laws that did not follow from those on the list, and moreover there would have been models of the listed laws that were not Boolean algebras.

This axiomatization is by no means the only one, or even necessarily the most natural given that we did not pay attention to whether some of the axioms followed from others but simply chose to stop when we noticed we had enough laws, treated further in the section on axiomatizations. Or the intermediate notion of axiom can be sidestepped altogether by defining a Boolean law directly as any tautology, understood as an equation that holds for all values of its variables over 0 and 1. All these definitions of Boolean algebra can be shown to be equivalent.

Duality Principle

Principle: If {X, R} is a poset, then {X, R(inverse)} is also a poset.

There is nothing magical about the choice of symbols for the values of Boolean algebra. We could rename 0 and 1 to say α and β, and as long as we did so consistently throughout it would still be Boolean algebra, albeit with some obvious cosmetic differences.

But suppose we rename 0 and 1 to 1 and 0 respectively. Then it would still be Boolean algebra, and moreover operating on the same values. However it would not be identical to our original Boolean algebra because now we find ∨ behaving the way ∧ used to do and vice versa. So there are still some cosmetic differences to show that we've been fiddling with the notation, despite the fact that we're still using 0s and 1s.

But if in addition to interchanging the names of the values we also interchange the names of the two binary operations, *now* there is no trace of what we have done. The end product is completely indistinguishable from what we started with. We might notice that the columns for $x \wedge y$ and $x \vee y$ in the truth tables had changed places, but that switch is immaterial.

When values and operations can be paired up in a way that leaves everything important unchanged when all pairs are switched simultaneously, we call the members of each pair dual to each other. Thus 0 and 1 are dual, and ∧ and ∨ are dual. The Duality Principle, also called De Morgan duality, asserts that Boolean algebra is unchanged when all dual pairs are interchanged.

One change we did not need to make as part of this interchange was to complement. We say that complement is a self-dual operation. The identity or do-nothing operation x (copy the input to the output) is also self-dual. A more complicated example of a self-dual operation is $(x \wedge y) \vee (y \wedge z) \vee (z \wedge x)$. There is no self-dual binary operation that depends on both its arguments. A composition of self-dual operations is a self-dual operation. For example, if $f(x, y, z) = (x \wedge y) \vee (y \wedge z) \vee (z \wedge x)$, then $f(f(x, y, z), x, t)$ is a self-dual operation of four arguments x,y,z,t.

The principle of duality can be explained from a group theory perspective by the fact that there are exactly four functions that are one-to-one mappings (automorphisms) of the set of Boolean polynomials back to itself: the identity function, the complement function, the dual function and the contradual function (complemented dual). These four functions form a group under function composition, isomorphic to the Klein four-group, acting on the set of Boolean polynomials. Walter Gottschalk remarked that consequently a more appropriate name for the phenomenon would be the *principle* (or *square*) *of quaternality*.

Diagrammatic Representations

Venn Diagrams

A Venn diagram is a representation of a Boolean operation using shaded overlapping regions. There is one region for each variable, all circular in the examples here. The interior and exterior of region x corresponds respectively to the values 1 (true) and 0 (false) for variable x. The shading indicates the value of the operation for each combination of regions, with dark denoting 1 and light 0 (some authors use the opposite convention).

The three Venn diagrams in the figure below represent respectively conjunction $x \wedge y$, disjunction $x \vee y$, and complement $\neg x$.

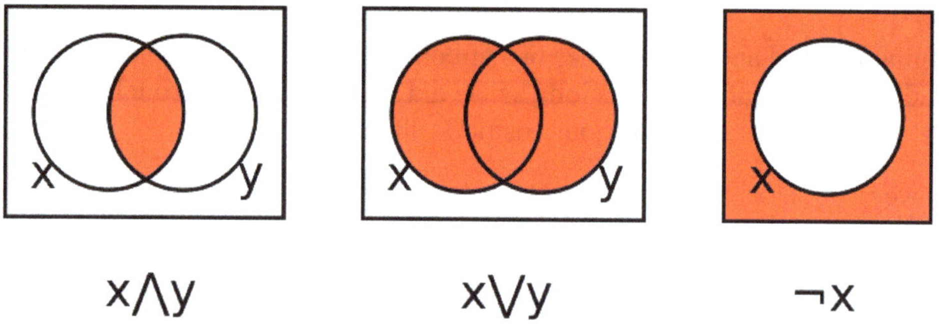

Figure. Venn diagrams for conjunction, disjunction, and complement.

For conjunction, the region inside both circles is shaded to indicate that $x \wedge y$ is 1 when both variables are 1. The other regions are left unshaded to indicate that $x \wedge y$ is 0 for the other three combinations.

The second diagram represents disjunction $x \vee y$ by shading those regions that lie inside either or both circles. The third diagram represents complement $\neg x$ by shading the region *not* inside the circle.

While we have not shown the Venn diagrams for the constants 0 and 1, they are trivial, being respectively a white box and a dark box, neither one containing a circle. However we could put a circle for x in those boxes, in which case each would denote a function of one argument, x, which returns the same value independently of x, called a constant function. As far as their outputs are concerned, constants and constant functions are indistinguishable; the difference is that a constant takes no arguments, called a *zeroary* or *nullary* operation, while a constant function takes one argument, which it ignores, and is a *unary* operation.

Venn diagrams are helpful in visualizing laws. The commutativity laws for $\wedge$ and $\vee$ can be seen from the symmetry of the diagrams: a binary operation that was not commutative would not have a symmetric diagram because interchanging x and y would have the effect of reflecting the diagram horizontally and any failure of commutativity would then appear as a failure of symmetry.

Idempotence of $\wedge$ and $\vee$ can be visualized by sliding the two circles together and noting that the shaded area then becomes the whole circle, for both $\wedge$ and $\vee$.

To see the first absorption law, $x \wedge (x \vee y) = x$, start with the diagram in the middle for $x \vee y$ and note that the portion of the shaded area in common with the x circle is the whole of the x circle. For the second absorption law, $x\vee(x\wedge y) = x$, start with the left diagram for $x\wedge y$ and note that shading the whole of the x circle results in just the x circle being shaded, since the previous shading was inside the x circle.

The double negation law can be seen by complementing the shading in the third diagram for $\neg x$, which shades the x circle.

To visualize the first De Morgan's law, $(\neg x)\wedge(\neg y) = \neg(x\vee y)$, start with the middle diagram for $x\vee y$ and complement its shading so that only the region outside both circles is shaded, which is what the right hand side of the law describes. The result is the same as if we shaded that region which is both outside the x circle *and* outside the y circle, i.e. the conjunction of their exteriors, which is what the left hand side of the law describes.

The second De Morgan's law, $(\neg x)\vee(\neg y) = \neg(x\wedge y)$, works the same way with the two diagrams interchanged.

The first complement law, $x\wedge\neg x = 0$, says that the interior and exterior of the x circle have no overlap. The second complement law, $x\vee\neg x = 1$, says that everything is either inside or outside the x circle.

Digital Logic Gates

Digital logic is the application of the Boolean algebra of 0 and 1 to electronic hardware consisting of logic gates connected to form a circuit diagram. Each gate implements a Boolean operation, and is depicted schematically by a shape indicating the operation. The shapes associated with the gates for conjunction (AND-gates), disjunction (OR-gates), and complement (inverters) are as follows.

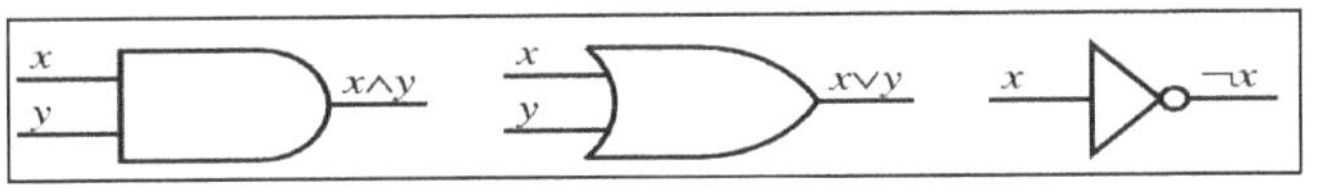

Logic gates.

The lines on the left of each gate represent input wires or *ports*. The value of the input is represented by a voltage on the lead. For so-called "active-high" logic, 0 is represented by a voltage close to zero or "ground", while 1 is represented by a voltage close to the supply voltage; active-low reverses this. The line on the right of each gate represents the output port, which normally follows the same voltage conventions as the input ports.

Complement is implemented with an inverter gate. The triangle denotes the operation that simply copies the input to the output; the small circle on the output denotes the actual inversion complementing the input. The convention of putting such a circle on any port means that the signal passing through this port is complemented on the way through, whether it is an input or output port.

The Duality Principle, or De Morgan's laws, can be understood as asserting that complementing all three ports of an AND gate converts it to an OR gate and vice versa, as shown in Figure below. Complementing both ports of an inverter however leaves the operation unchanged.

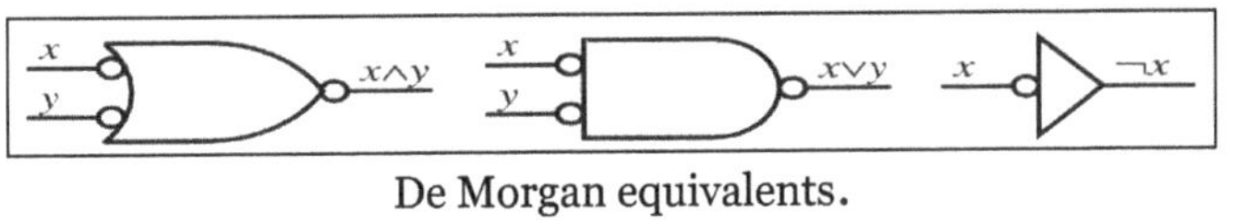

De Morgan equivalents.

More generally one may complement any of the eight subsets of the three ports of either an AND or OR gate. The resulting sixteen possibilities give rise to only eight Boolean operations, namely those with an odd number of 1's in their truth table. There are eight such because the "odd-bit-out" can be either 0 or 1 and can go in any of four positions in the truth table. There being sixteen binary Boolean operations, this must leave eight operations with an even number of 1's in their truth tables. Two of these are the constants 0 and 1 (as binary operations that ignore both their inputs); four are the operations that depend nontrivially on exactly one of their two inputs, namely x, y, $\neg x$, and $\neg y$; and the remaining two are $x \oplus y$ (XOR) and its complement $x \equiv y$.

Propositional Logic

Propositional logic is a logical system that is intimately connected to Boolean algebra. Many syntactic concepts of Boolean algebra carry over to propositional logic with only minor changes in notation and terminology, while the semantics of propositional logic are defined via Boolean algebras in a way that the tautologies (theorems) of propositional logic correspond to equational theorems of Boolean algebra.

Syntactically, every Boolean term corresponds to a propositional formula of propositional logic. In this translation between Boolean algebra and propositional logic, Boolean variables x,y... become propositional variables (or atoms) P,Q,..., Boolean terms such as $x \vee y$ become propositional formulas $P \vee Q$, 0 becomes *false* or $\bot$, and 1 becomes *true* or T. It is convenient when referring to generic propositions to use Greek letters Φ, Ψ,... as metavariables (variables outside the language of propositional calculus, used when talking *about* propositional calculus) to denote propositions.

The semantics of propositional logic rely on truth assignments. The essential idea of a truth assignment is that the propositional variables are mapped to elements of a fixed Boolean algebra, and then the truth value of a propositional formula using these letters is the element of the Boolean algebra that is obtained by computing the value of the Boolean term corresponding to the formula. In classical semantics, only the two-element Boolean algebra is used, while in Boolean-valued semantics arbitrary Boolean algebras are considered. A tautology is a propositional formula that is assigned truth value *1* by every truth assignment of its propositional variables to an arbitrary Boolean algebra (or, equivalently, every truth assignment to the two element Boolean algebra).

These semantics permit a translation between tautologies of propositional logic and equational theorems of Boolean algebra. Every tautology Φ of propositional logic can be expressed as the Boolean equation $\Phi = 1$, which will be a theorem of Boolean algebra. Conversely every theorem $\Phi = \Psi$ of Boolean algebra corresponds to the tautologies $(\Phi \vee \neg\Psi) \wedge (\neg\Phi \vee \Psi)$ and $(\Phi \wedge \Psi) \vee (\neg\Phi \wedge \neg\Psi)$. If $\rightarrow$ is in the language these last tautologies can also be written as $(\Phi \rightarrow \Psi) \wedge (\Psi \rightarrow \Phi)$, or as two separate theorems $\Phi \rightarrow \Psi$ and $\Psi \rightarrow \Phi$; if $\equiv$ is available then the single tautology $\Phi \equiv \Psi$ can be used.

Applications

One motivating application of propositional calculus is the analysis of propositions and deductive arguments in natural language. Whereas the proposition "if $x = 3$ then $x+1 = 4$" depends on the meanings of such symbols as + and 1, the proposition "if $x = 3$ then $x = 3$" does not; it is true merely by virtue of its structure, and remains true whether "$x = 3$" is replaced by "$x = 4$" or "the moon is made of green cheese." The generic or abstract form of this tautology is "if P then P", or in the language of Boolean algebra, "$P \to P$".

Replacing P by $x = 3$ or any other proposition is called instantiation of P by that proposition. The result of instantiating P in an abstract proposition is called an instance of the proposition. Thus "$x = 3 \to x = 3$" is a tautology by virtue of being an instance of the abstract tautology "$P \to P$". All occurrences of the instantiated variable must be instantiated with the same proposition, to avoid such nonsense as $P \to x = 3$ or $x = 3 \to x = 4$.

Propositional calculus restricts attention to abstract propositions, those built up from propositional variables using Boolean operations. Instantiation is still possible within propositional calculus, but only by instantiating propositional variables by abstract propositions, such as instantiating Q by $Q \to P$ in $P \to (Q \to P)$ to yield the instance $P \to ((Q \to P) \to P)$.

(The availability of instantiation as part of the machinery of propositional calculus avoids the need for metavariables within the language of propositional calculus, since ordinary propositional variables can be considered within the language to denote arbitrary propositions. The metavariables themselves are outside the reach of instantiation, not being part of the language of propositional calculus but rather part of the same language for talking about it that this sentence is written in, where we need to be able to distinguish propositional variables and their instantiations as being distinct syntactic entities.)

Deductive Systems for Propositional Logic

An axiomatization of propositional calculus is a set of tautologies called axioms and one or more inference rules for producing new tautologies from old. A *proof* in an axiom system A is a finite nonempty sequence of propositions each of which is either an instance of an axiom of A or follows by some rule of A from propositions appearing earlier in the proof (thereby disallowing circular reasoning). The last proposition is the theorem proved by the proof. Every nonempty initial segment of a proof is itself a proof, whence every proposition in a proof is itself a theorem. An axiomatization is sound when every theorem is a tautology, and complete when every tautology is a theorem.

Sequent Calculus

Propositional calculus is commonly organized as a Hilbert system, whose operations are just those of Boolean algebra and whose theorems are Boolean tautologies, those Boolean terms equal to the Boolean constant 1. Another form is sequent calculus, which has two sorts, propositions as in ordinary propositional calculus, and pairs of lists of propositions called sequents, such as $A \vee B, A \wedge C, \ldots \vdash A, B \to C, \ldots$. The two halves of a sequent are called the antecedent and the succedent respectively. The customary metavariable denoting an antecedent or part thereof is Γ, and for a succedent Δ; thus $\Gamma, A \vdash \Delta$ would denote a sequent whose succedent is a list Δ and whose antecedent is a list Γ with an additional proposition A appended after it. The antecedent is interpreted as the conjunction of its propositions, the succedent as the disjunction of its propositions, and the sequent itself as the entailment of the succedent by the antecedent.

Entailment differs from implication in that whereas the latter is a binary *operation* that returns a value in a Boolean algebra, the former is a binary *relation* which either holds or does not hold. In this sense entailment is an *external* form of implication, meaning external to the Boolean algebra, thinking of the reader of the sequent as also being external and interpreting and comparing antecedents and succedents in some Boolean algebra. The natural interpretation of $\vdash$ is as $\leq$ in the partial order of the Boolean algebra defined by $x \leq y$ just when $x \vee y = y$. This ability to mix external implication $\vdash$ and internal implication $\rightarrow$ in the one logic is among the essential differences between sequent calculus and propositional calculus.

Applications

Boolean algebra as the calculus of two values is fundamental to computer circuits, computer programming, and mathematical logic, and is also used in other areas of mathematics such as set theory and statistics.

Computers

In the early 20th century, several electrical engineers intuitively recognized that Boolean algebra was analogous to the behavior of certain types of electrical circuits. Claude Shannon formally proved such behavior was logically equivalent to Boolean algebra in his 1937 master's thesis, *A Symbolic Analysis of Relay and Switching Circuits.*

Today, all modern general purpose computers perform their functions using two-value Boolean logic; that is, their electrical circuits are a physical manifestation of two-value Boolean logic. They achieve this in various ways: as voltages on wires in high-speed circuits and capacitive storage devices, as orientations of a magnetic domain in ferromagnetic storage devices, as holes in punched cards or paper tape, and so on. (Some early computers used decimal circuits or mechanisms instead of two-valued logic circuits.)

Of course, it is possible to code more than two symbols in any given medium. For example, one might use respectively 0, 1, 2, and 3 volts to code a four-symbol alphabet on a wire, or holes of different sizes in a punched card. In practice, the tight constraints of high speed, small size, and low power combine to make noise a major factor. This makes it hard to distinguish between symbols when there are several possible symbols that could occur at a single site. Rather than attempting to distinguish between four voltages on one wire, digital designers have settled on two voltages per wire, high and low.

Computers use two-value Boolean circuits for the above reasons. The most common computer architectures use ordered sequences of Boolean values, called bits, of 32 or 64 values, e.g. 011010001 10101100101010101001011. When programming in machine code, assembly language, and certain other programming languages, programmers work with the low-level digital structure of the data registers. These registers operate on voltages, where zero volts represents Boolean 0, and a reference voltage (often +5V, +3.3V, +1.8V) represents Boolean 1. Such languages support both numeric operations and logical operations. In this context, "numeric" means that the computer treats sequences of bits as binary numbers (base two numbers) and executes arithmetic operations like add, subtract, multiply, or divide. "Logical" refers to the Boolean logical operations of disjunction, conjunction, and negation between two sequences of bits, in which each bit in one sequence is simply compared to its counterpart in the other sequence. Programmers therefore have the option of working in and applying the rules of either numeric algebra or Boolean algebra as needed. A core differentiating feature between these families of operations is the existence of the carry operation in the first but not the second.

Two-valued Logic

Other areas where two values is a good choice are the law and mathematics. In everyday relaxed conversation, nuanced or complex answers such as "maybe" or "only on the weekend" are acceptable. In more focused situations such as a court of law or theorem-based mathematics however it is deemed advantageous to frame questions so as to admit a simple yes-or-no answer—is the defendant guilty or not guilty, is the proposition true or false—and to disallow any other answer. However much of a straitjacket this might prove in practice for the respondent, the principle of the simple yes-no question has become a central feature of both judicial and mathematical logic, making two-valued logic deserving of organization and study in its own right.

A central concept of set theory is membership. Now an organization may permit multiple degrees of membership, such as novice, associate, and full. With sets however an element is either in or out. The candidates for membership in a set work just like the wires in a digital computer: each candidate is either a member or a nonmember, just as each wire is either high or low.

Algebra being a fundamental tool in any area amenable to mathematical treatment, these considerations combine to make the algebra of two values of fundamental importance to computer hardware, mathematical logic, and set theory.

Two-valued logic can be extended to multi-valued logic, notably by replacing the Boolean domain {0, 1} with the unit interval [0,1], in which case rather than only taking values 0 or 1, any value between and including 0 and 1 can be assumed. Algebraically, negation (NOT) is replaced with $1 - x$, conjunction (AND) is replaced with multiplication (xy), and disjunction (OR) is defined via De Morgan's law. Interpreting these values as logical truth values yields a multi-valued logic, which forms the basis for fuzzy logic and probabilistic logic. In these interpretations, a value is interpreted as the "degree" of truth – to what extent a proposition is true, or the probability that the proposition is true.

Boolean Operations

The original application for Boolean operations was mathematical logic, where it combines the truth values, true or false, of individual formulas.

Natural languages such as English have words for several Boolean operations, in particular conjunction (*and*), disjunction (*or*), negation (*not*), and implication (*implies*). *But not* is synonymous with *and not*. When used to combine situational assertions such as "the block is on the table" and "cats drink milk," which naively are either true or false, the meanings of these logical connectives often have the meaning of their logical counterparts. However, with descriptions of behavior such as "Jim walked through the door", one starts to notice differences such as failure of commutativity, for example the conjunction of "Jim opened the door" with "Jim walked through the door" in that order is not equivalent to their conjunction in the other order, since *and* usually means *and then* in such cases. Questions can be similar: the order "Is the sky blue, and why is the sky blue?" makes more sense than the reverse order. Conjunctive commands about behavior are like behavioral assertions, as in *get dressed and go to school.* Disjunctive commands such *love me or leave me* or *fish or cut bait* tend to be asymmetric via the implication that one alternative is less preferable. Conjoined nouns such as *tea and milk* generally describe aggregation as with set union while *tea or milk* is a choice. However context can reverse these senses, as in *your choices are coffee and tea* which usually means the same as *your choices are coffee or tea* (alternatives). Double negation as in "I don't not like milk" rarely means literally "I do like milk" but rather conveys some sort of hedging, as though to imply that there is a third possibility. "Not not P" can be loosely interpreted as "surely P", and although *P* necessarily implies "not not *P*" the converse is suspect in English, much as with intuitionistic logic. In view of the highly idiosyncratic usage of conjunctions in natural languages, Boolean algebra cannot be considered a reliable framework for interpreting them.

Boolean operations are used in digital logic to combine the bits carried on individual wires, thereby interpreting them over {0,1}. When a vector of n identical binary gates are used to combine two bit vectors each of n bits, the individual bit operations can be understood collectively as a single operation on values from a Boolean algebra with 2^n elements.

Naive set theory interprets Boolean operations as acting on subsets of a given set X. As we saw earlier this behavior exactly parallels the coordinate-wise combinations of bit vectors, with the union of two sets corresponding to the disjunction of two bit vectors and so on.

The 256-element free Boolean algebra on three generators is deployed in computer displays based on raster graphics, which use bit blit to manipulate whole regions consisting of pixels, relying on Boolean operations to specify how the source region should be combined with the destination, typically with the help of a third region called the mask. Modern video cards offer all 2^{23} = 256 ternary operations for this purpose, with the choice of operation being a one-byte (8-bit) parameter. The constants SRC = 0xaa or 10101010, DST = 0xcc or 11001100, and MSK = 0xf0 or 11110000 allow Boolean operations such as (SRC^DST)&MSK (meaning XOR the source and destination and then AND the result with the mask) to be written directly as a constant denoting a byte calculated at compile time, 0x60 in the (SRC^DST)&MSK example, 0x66 if just SRC^DST, etc. At run time the video card interprets the byte as the raster operation indicated by the original expression in a uniform way that requires remarkably little hardware and which takes time completely independent of the complexity of the expression.

Solid modeling systems for computer aided design offer a variety of methods for building objects from other objects, combination by Boolean operations being one of them. In this method the space in which objects exist is understood as a set S of voxels (the three-dimensional analogue of pixels in two-dimensional graphics) and shapes are defined as subsets of S, allowing objects to be combined as sets via union, intersection, etc. One obvious use is in building a complex shape from simple shapes simply as the union of the latter. Another use is in sculpting understood as removal of material: any grinding, milling, routing, or drilling operation that can be performed with physical machinery on physical materials can be simulated on the computer with the Boolean operation $x \wedge \neg y$ or $x - y$, which in set theory is set difference, remove the elements of y from those of x. Thus given two shapes one to be machined and the other the material to be removed, the result of machining the former to remove the latter is described simply as their set difference.

Boolean Searches

Search engine queries also employ Boolean logic. For this application, each web page on the Internet may be considered to be an "element" of a "set". The following examples use a syntax supported by Google.

- Doublequotes are used to combine whitespace-separated words into a single search term.
- Whitespace is used to specify logical AND, as it is the default operator for joining search terms:

"Search term 1" "Search term 2"

- The OR keyword is used for logical OR:

"Search term 1" OR "Search term 2"

- A prefixed minus sign is used for logical NOT:

"Search term 1" –"Search term 2"

Boolean Algebra Examples

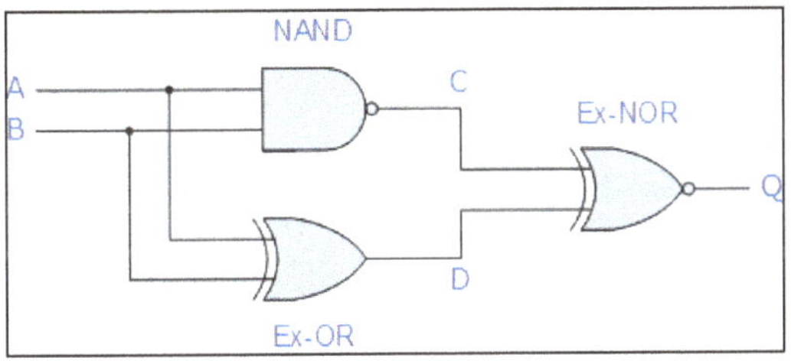

Boolean Algebra can be used to identify unnecessary logic gates within a digital logic design reducing the number of gates required.

Boolean Algebra Example 1:

Construct a Truth Table for the logical functions at points C, D and Q in the following circuit and identify a single logic gate that can be used to replace the whole circuit.

First observations tell us that the circuit consists of a 2-input NAND gate, a 2-input EX-OR gate and finally a 2-input EX-NOR gate at the output. As there are only 2 inputs to the circuit labelled A and B, there can only be 4 possible combinations of the input (2^2) and these are: 0-0, 0-1, 1-0 and finally 1-1. Plotting the logical functions from each gate in tabular form will give us the following truth table for the whole of the logic circuit below.

Inputs		Output at		
A	B	C	D	Q
0	0	1	0	0
0	1	1	1	1
1	0	1	1	1
1	1	0	0	1

From the truth table above, column C represents the output function generated by the NAND gate, while column D represents the output function from the Ex-OR gate. Both of these two output expressions then become the input condition for the Ex-NOR gate at the output.

It can be seen from the truth table that an output at Q is present when any of the two inputs A or B are at logic 1. The only truth table that satisfies this condition is that of an OR Gate. Therefore, the whole of the above circuit can be replaced by just one single 2-input OR Gate.

Boolean Algebra Example 2:

Find the Boolean algebra expression for the following system.

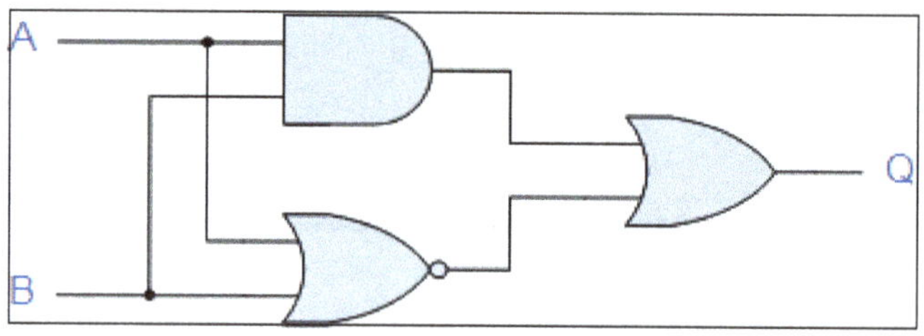

The system consists of an AND Gate, a NOR Gate and finally an OR Gate. The expression for the AND gate is A.B, and the expression for the NOR gate is $\overline{A+B}$. Both these expressions are also separate inputs to the OR gate which is defined as A+B. Thus the final output expression is given as:

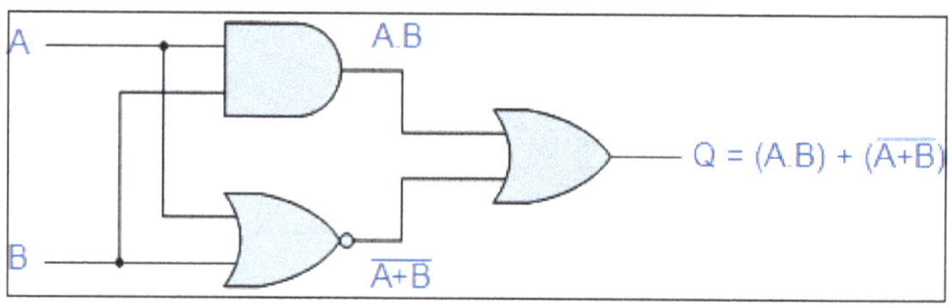

The output of the system is given as Q = (A.B) + ($\overline{A+B}$), but the notation $\overline{A+B}$ is the same as the De Morgan´s notation $\overline{A}.\overline{B}$, Then substituting $\overline{A}.\overline{B}$ into the output expression gives us a final output notation of Q = (A.B)+($\overline{A}.\overline{B}$), which is the Boolean notation for an Exclusive-NOR Gate as seen in the previous topic.

Inputs		Intermediates		Output
B	A	A.B	$\overline{A+B}$	Q
0	0	0	1	1
0	1	0	0	0
1	0	0	0	0
1	1	1	0	1

Then, the whole circuit above can be replaced by just one single Exclusive-NOR Gate and indeed an Exclusive-NOR Gate is made up of these individual gate functions.

Boolean Algebra Example 3:

Find the Boolean algebra expression for the following system.

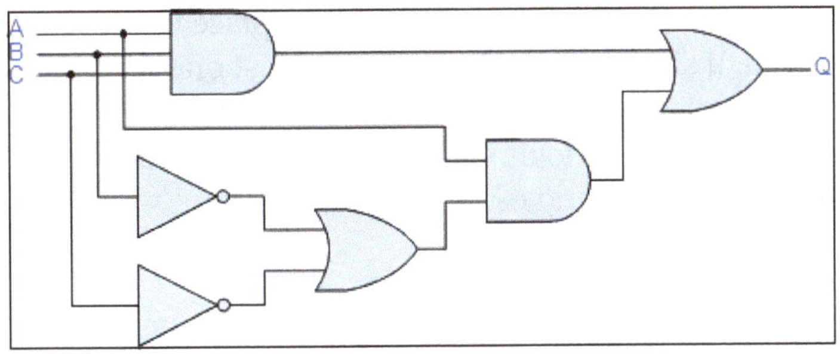

This system may look more complicated than the other two to analyse but again, the logic circuit just consists of simple AND, OR and NOT gates connected together.

As with the previous Boolean examples, we can simplify the circuit by writing down the Boolean notation for each logic gate function in turn in order to give us a final expression for the output at Q.

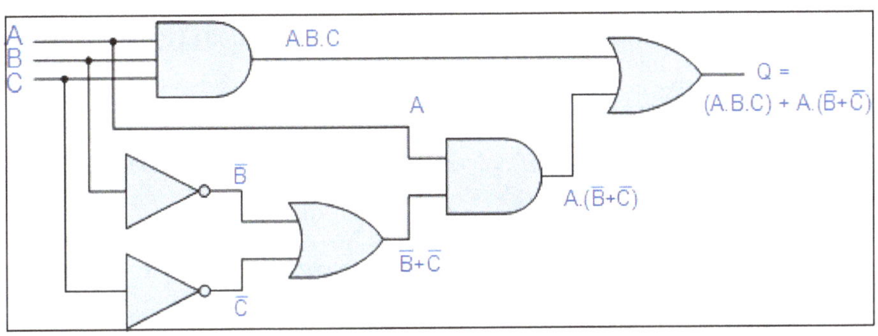

The output from the 3-input AND gate is only at logic "1" when ALL the gates inputs are HIGH at logic level "1" (A.B.C). The output from the lower OR gate is only a "1" when one or both inputs B or C are at logic level "0". The output from the 2-input AND gate is a "1" when input A is a "1" and inputs B or C are at "0". Then the output at Q is only a "1" when inputs A.B.C equal "1" or A is equal to "1" and both inputs B or C equal "0", $A.(\overline{B}+\overline{C})$.

By using "de Morgan's theorem" inputs B and input C cancel out as to produce an output at Q they can be either at logic "1" or at logic "0". Then this just leaves input A as the only input needed to give an output at Q as shown in the table below.

Inputs			Intermediates					Output
C	B	A	A.B.C	B	C	B+C	A.(B+C)	Q
0	0	0	0	1	1	1	0	0
0	0	1	0	1	1	1	1	1
0	1	0	0	0	1	1	0	0
0	1	1	0	0	1	1	1	1
1	0	0	0	1	0	1	0	0
1	0	1	0	1	0	1	1	1
1	1	0	0	0	0	0	0	0
1	1	1	1	0	0	0	0	1

Then we can see that the entire logic circuit above can be replaced by just one single input labelled A thereby reducing a circuit of six individual logic gates to just one single piece of wire, (or Buffer). This type of circuit analysis using Boolean Algebra can be very powerful and quickly identify any unnecessary logic gates within a digital logic design thereby reducing the number of gates required, the power consumption of the circuit and of course the cost.

References

- Digital-Electronics-Introduction-Number-Systems-Number-Codes: engineersgarage.com, Retrieved 22 June 2020
- "National Inventors Hall of Fame – George R. Stibitz". 20 August 2008. Archived from the original on 8 July 2010. Retrieved 5 July 2020
- J.E.H. Smith (2008). Leibniz: What Kind of Rationalist?: What Kind of Rationalist?. Springer. p. 415. ISBN 978-1-4020-8668-7
- Write-your-name-in-binary-code, educational-resources: sciencefriday.com, Retrieved 25 July 2020
- "Pioneers – The people and ideas that made a difference – George Stibitz (1904–1995)". Kerry Redshaw. 20 February 2006. Retrieved 5 July 2020
- Binary-codes, computer-logical-organization: tutorialspoint.com, Retrieved 15 April 2020
- J. Michael Dunn; Gary M. Hardegree (2001). Algebraic methods in philosophical logic. Oxford University Press US. p. 2. ISBN 978-0-19-853192-0
- Robson, Eleanor; Stedall, Jacqueline, eds. (2009), "Myth No. 2: the Horus eye fractions", The Oxford Handbook of the History of Mathematics, Oxford University Press, p. 790, ISBN 9780199213122
- Binary-coded-decimal: electronics-tutorials.ws, Retrieved 25 May 2020
- Bender, Andrea; Beller, Sieghard (16 December 2013). "Mangarevan invention of binary steps for easier calculation". Proceedings of the National Academy of Sciences. 111: 1322–1327. doi:10.1073/pnas.1309160110. PMC 3910603. PMID 24344278

5

Diverse Aspects of Digital Electronics

Some of the concepts which are studied within digital electronics are: delays and hazards, Boolean algebra, Karnaugh maps, encoder, decoder, multiplexers and demultiplexers, Arithmetic Logic Unit (ALU) and digital system design. This chapter discusses in detail all these concepts.

Delays and Hazards

The term "hazards" specifies the unwanted switching transient, false output or glitches which appear at the output of a circuit. Hazards in any system are an undesirable effect caused by either internal deficiency of the system or external disturbances. In combinational circuits, these transients or glitches are due to finite propagation delay of the components along the different paths within the network. The difference in propagation delays are corresponding to different signal paths results in hazards.

The hazard can also be defined as the actual or potential malfunction of a logic circuit during transition between two input states when single variable changes. The term malfunction means any deviation from the intended response. These logic hazards have the same basic problem. When the changes in the input signal do not change the output signal due to some form of delay caused by logic building blocks like NOT, AND OR gates etc. This generates a malfunctioning in the performance of the circuit. This is however a temporary problem and the logic will finally generate a desired function output. Even though the combinational logic system is generating a correct output after some time, it is essential that hazards be eliminated as they can have an effect on the overall performance of the system.

In the combinational circuits, the hazards will produce an unwanted output value. If such combinational circuits are used as building blocks of bigger digital system like asynchronous sequential circuit then it will produce a transition to incorrect state table.

Types of Hazards

The hazards result in the false output value when different paths from input to output have different propagation delays. It is the deviation from the standard response. There are three types of hazards: Static hazards, dynamic hazards and function hazards as shown in figure.

- In static hazards, when one input variable changes, the output changes momentarily before stabilizing to the correct values. There are two types of static hazards:
    - Static 0 hazard: A static 0 hazard takes place when the output momentarily goes to 1 (HIGH) for a short duration when it should remain at a constant value 0 (LOW).

- Static 1 hazard: If the output is ideally supposed to be 1 (HIGH) but due to difference in propagation delay, if it turns out to be 0 (LOW) for short period of time, then the circuit said to have static 1 hazard.

- Dynamic hazard: In static hazard, the output changes only twice (1→0→1 or 0→1→0). In dynamic hazards, there is a possibility of an output changing more than once as a result of a single input change. When an output is supposed to change value from 0 to 1 or 1 to 0, this output may change 3 or more times. If this situation appears at output, the circuit is said to have a dynamic hazards.
- Function hazards: A hazard that can be caused in the output signal, when more than one input signal is changed simultaneously, is called a function hazard.

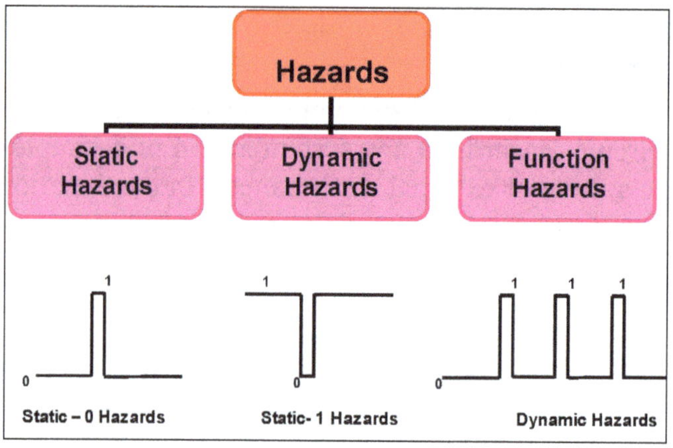

Static Hazards

A static hazard occurs when a single input variable change should cause no change in the output of combinational circuit but a short glitch of the incorrect logic level occurs. Here, the output changes momentarily when it should not. The static-0 a hazard is also called POS hazards and static-1 hazard is also called SOP hazards.

A glitch is an undesired transition that occurs before the signal settles to its intended value. In other terms, glitch is an electric pulse of short duration that is usually the result of a fault or design error in a digital circuit. The glitches are generated because of the propagation delays of the logic gates. Propagation delay is a time associated with any digital circuit and is the time between when an input to the circuit changes until that change propagates through the circuit and changes the output. Every logic gate has its own propagation delay and is dependent on the fabrication technology of the device. Figure below indicates the propagation delay in the NOT gate.

A static hazard in a logic network is a transient change of an output value which is supposed to remain fixed during the transition between two input states differing in the value of only one variable. When the output is supposed to remain at static condition "0", but goes to logic 1 momentarily during the transition of input states, then the hazard is called as static 0 hazard. On the other hand if a momentary "0" is likely to occur in the output when it is expected to remain in static condition "1", the hazard is called as static 1 hazard.

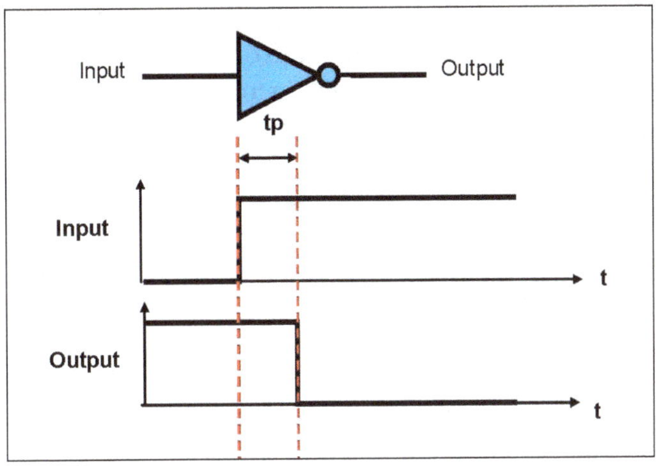

Propagation delay in NOT gate.

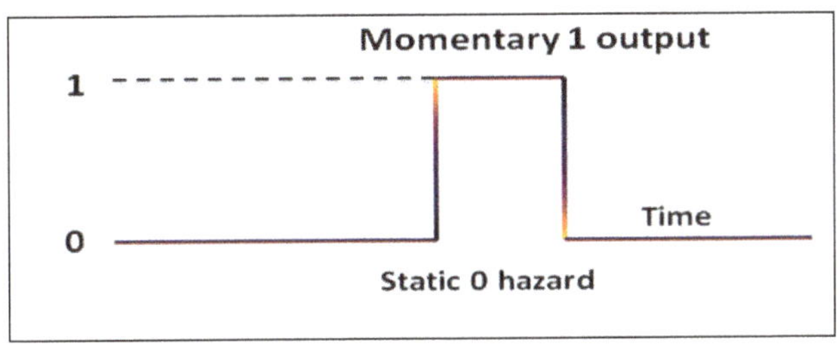

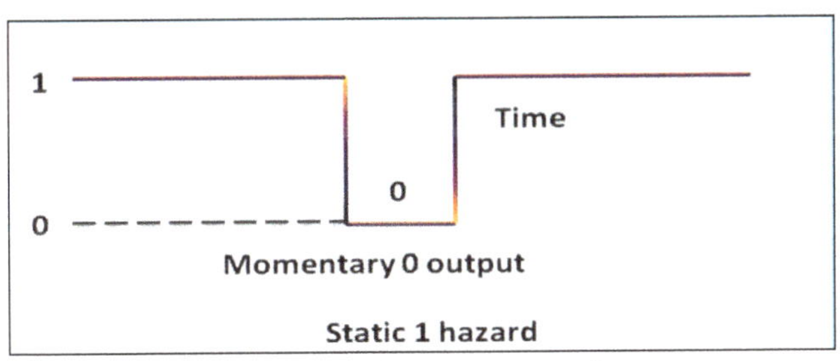

Static-0 and Static-1 Hazards.

Basic Static-0 Hazards Circuit

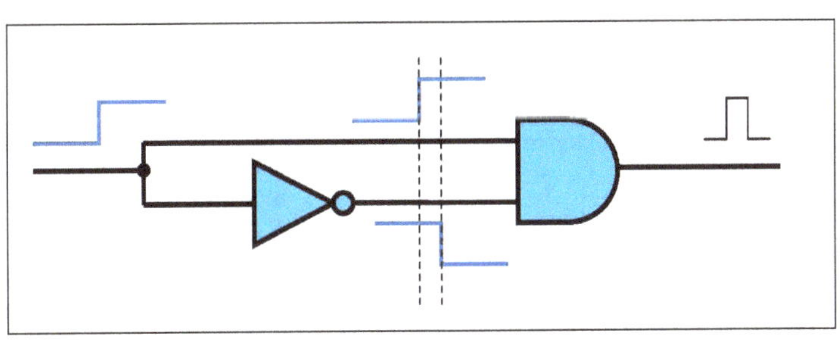

Basic static-0 hazard circuit.

A simple circuit to demonstrate the static 0 hazards is shown in figure. This simple circuit consist of an AND gate and a NOT gate. The input signal is applied directly to one of the inputs of the AND gate and the other input of the AND gate is obtained through the inverter i.e. NOT gate. As the NOT gate is present at one of the inputs of AND gate the signal at that input is delayed by one NOT gate delay time.

When the input signal is at logic 0, the NOT gate output is at logic 1 hence the output is at logic 0. When the input changes the state from 0 to 1 then the output of AND gate produces a small glitch which goes HIGH(1) momentarily and returns back to logic 0 after some delay. The duration of the glitch or small pulse is proportional to the delay of the NOT gate.

Combinational Circuit with Static-0 Hazard

A static hazard occurs when one or more combinational logic circuit input signals change value, and the final steady state of output value remains the same. Let us now study the hazards in any combinational circuit in general. To demonstrate the hazards in general combinational circuit consider the circuit shown in figure. The circuit consist of 3 AND gates and a NOT gate along with their interconnections. The figure also indicates the waveforms at various points in the circuit and the glitch generated at the output.

Let us assume A=0 and B=1, the output of two AND gates is G1=0 and G2=1. At t=t1, an input variable A changes its state form 0 to become 1. So now both the inputs are A=B=1. The output of gate 1 i.e. G1 becomes high after a time delay equal to the propagation delay of AND gate 1. The output of gate 2 i.e. G2 goes low after a time delay which is equal to the sum of propagation delays of the inverter and the AND gate 2. Hence output of the final AND gate, Y goes high for a short duration. After some time the output turns back to logic 0 as expected. This short duration HIGH pulse is unwanted and called as a static 0 hazard.

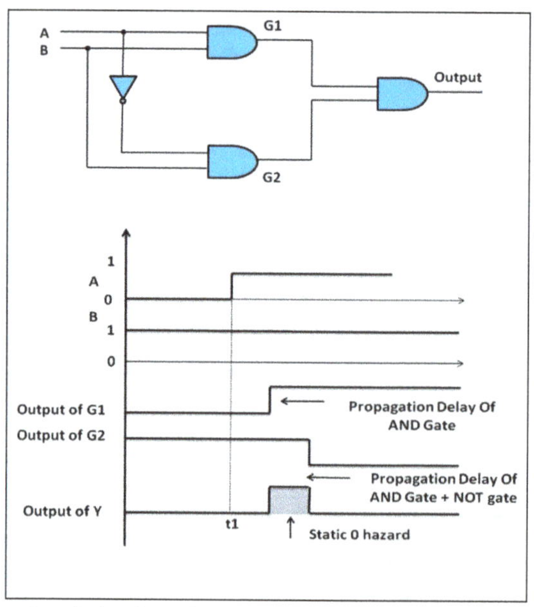

A combinational circuit with static-0 hazard and relevant waveforms.

Basic Static-1 Hazard Circuit

A simple circuit to demonstrate the static 1 hazard is shown in figure below. This simple circuit consist of an OR gate and a NOT gate. The input signal is applied directly to one of the inputs of the OR gate and the other input of the OR gate is obtained through the inverter i.e. NOT gate. As the NOT gate is present at one of the inputs of OR gate the signal at that input is delayed by one NOT gate delay time.

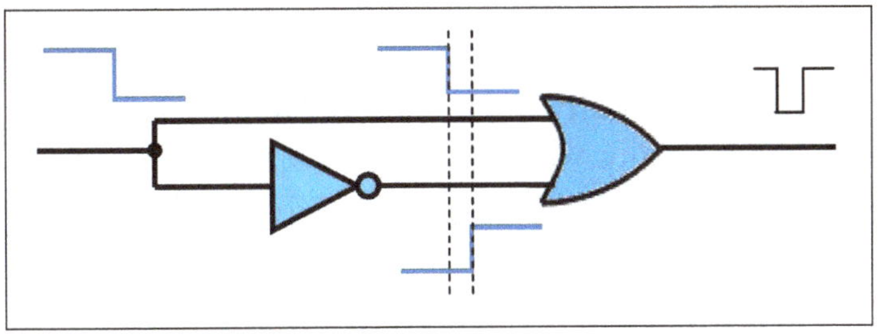

When the input signal is at logic 1, the NOT gate output is at logic 0, hence the output is at logic 1. When the input changes the state from 1 to 0 then the output of OR gate produces a small glitch which goes momentarily 0 (LOW). The duration of the glitch or small pulse is proportional to the delay of the NOT gate and then the output returns back to logic 1.

Combinational Circuit with Static-1 Hazard

A static hazard occurs when one or more combinational logic circuit input signals change value, and the final steady state of output value remains the same. Let us now study the hazards in any combinational circuit in general. To demonstrate the hazards in general combinational circuit consider the circuit shown in figure below. The circuit consist of 3 AND gates and a NOT gate along with their interconnections. This figure also indicates the waveforms at various points in the circuit and the glitch generated at the output.

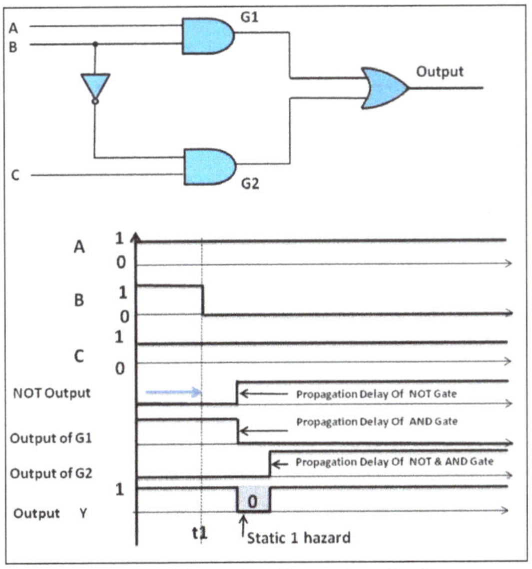

Combinational circuit with static 1 hazard and relevant waveforms.

Let us consider A=1, B=1 and C=1 as initial condition. The output of two AND gates is G1=1 and G2=0 and the final output Y is 1. At time t=t1, an input variable B changes its state form 1 to become 0. Due to this change in the input condition, the gate outputs will also change. The G1 gate goes from 1 to 0 after finite propagation delay and the output of G2 gate changes the output from 0 to 1 only after the propagation delay of NOT gate and AND gate delay. During this period, both gates G1 and G2 are at logic 0. Thus the output Y goes to 0 momentarily as shown in figure. This short duration LOW pulse is unwanted and called as static -1 hazard. Here we have assumed the propagation delays of AND and NOT gates are almost similar. The output returns back to logic 1 later on as expected.

Dynamic Hazards

In dynamic hazards, there is a possibility of an output changing more than once as a result of a single input change. In a larger circuit, whenever there are different routes from inputs to output then there is a possibility of dynamic hazards. When an output is supposed to change value from 0 to 1 or 1 to 0, this output may change 3 or more times. If this situation appears at output, the circuit is said to have a dynamic hazards.

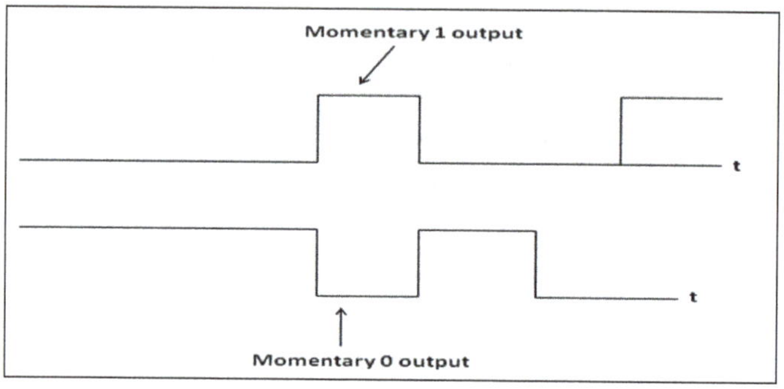

When multiple glitches occur at the output from multilevel circuit because of changing inputs, the only reason for this could be existence of three or more asymmetric paths (delay-wise) of that input to the output. In such a circuit, routes may have different delays; it is a potential dynamic hazard. Here changing output values differ from the expected output. For example, a logic circuit is expected to change from 1 to 0, but instead it changes from 1 to 0 then 1 and finally switches to the correct value 0 as shown in figure.

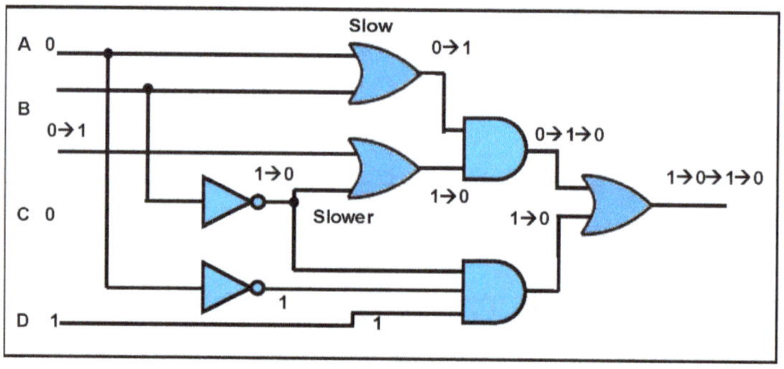

Example of Dynamic Hazards.

Dynamic hazards exist when there are multiple paths with different delays from the changing input to the changing output. Dynamic hazards do not occur in properly designed two level ANDOR or OR-AND circuits. It should be noted that a two level AND-OR or OR-AND circuit is properly designed if a variable and its complement are never input to the same first level gate.

Function Hazards

In the logic circuits, when more than one input variable changes at the same time, non-solvable hazards occurs. These non-solvable hazards are known as function hazards. Function hazards cannot be eliminated as the actual problem lies with the specification of the Boolean function. The only way to avoid the function hazards is to restrict the changing of input variables. Ensure that only one input changes at a given time. In reality, restrictions are not always possible. For example consider a circuit with three inputs. One of the inputs is connected to a clock circuit, the other one is tied to 1 or 0 and the third one is connected to random noise source. In such a case, it is impossible to restrict the other inputs from changing. Thus restricting the changing inputs to only one is not an effective solution.

Static Hazards Detection and Elimination

Hazards are unwanted switching transients that may appear at the output of a logic circuit because of different paths exhibit different propagation delays. The effect of seriousness of hazards is dependent on the devices or systems which receive the momentary transient signals. In case of digital systems the presence of hazards can cause the system to malfunction. Hazards can occur in combinational circuit as well as in sequential circuits. To detect and eliminate the static hazards from logic circuit following procedure is adopted:

- Write the Boolean expression for the given circuit and convert it into SOP form.
- Draw the k-map for the circuit.
- Adjacent 1's (or 0s if the circuit is in POS form) not covered by the same prime implicant indicates a static hazard exist.
- Remove the hazard by adding redundant prime implicants so that any two adjacent 1's (or 0's if the circuit is in POS form) are covered by the same prime implicant.
- Write the Boolean expression and redraw the logic circuit.

Elimination of Static Hazards

To eliminate the static hazards, one has to use additional logic to provide the holding action which is necessary for eliminating the static hazards. This additional logic is determined by grouping the two adjacent 1's or 0's which are responsible for static hazard, into a common subcube. Let us consider a simple example to demonstrate this.

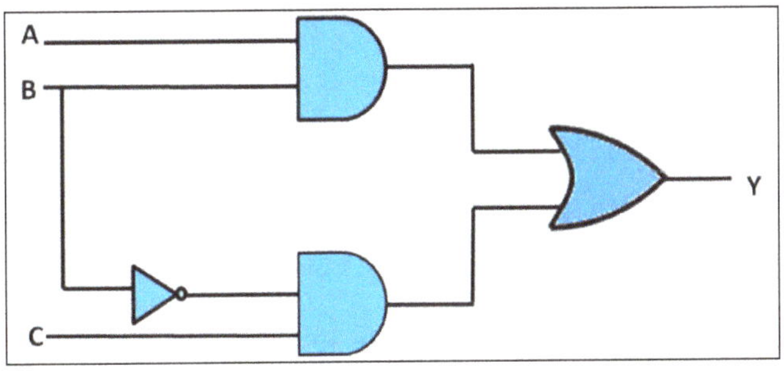

Logic circuit for demonstrating elimination of hazards.

Step 1: Write the Boolean expression for the given circuit and convert it into SOP form. Here expansion is carried out so as to include all the minterms.

$$Y = AB + B'C$$
$$= AB(C + C') + B'C(A + A') = ABC + ABC' + AB'C + A'B'C$$

Step-2: Draw the k-map.

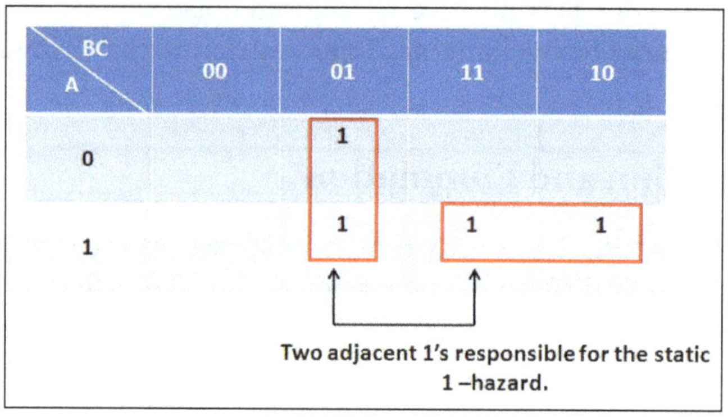

Step 3: Adjacent 1's not covered by the same prime implicant indicates a static hazard exist.

Step-4: Remove the hazard by adding redundant prime implicants so that any two adjacent 1's are covered by the same prime implicant. Group the two adjacent 1's responsible for the static-1 hazard into another pair.

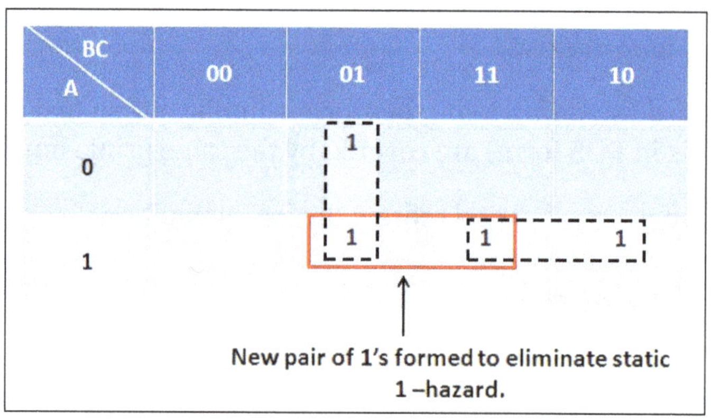

Step 5: Realize the new hazard free circuit.

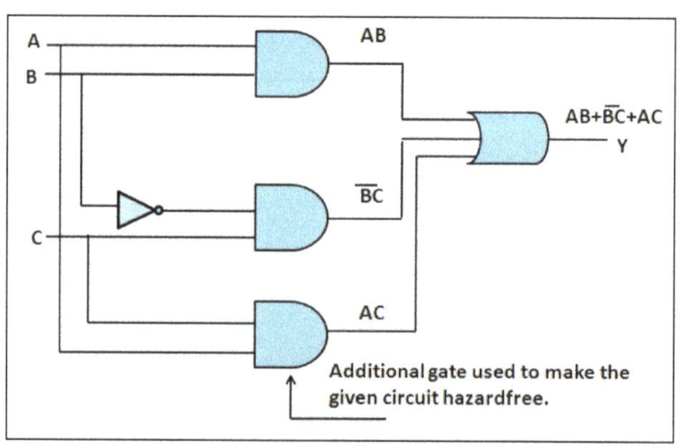

The new output for output Y after the formation of new pair is as follows:

$$Y = AB + B'C + AC \leftarrow \text{New term}$$

This provides a new hazard free circuit.

Boolean Algebra

Boolean algebra is an algebra for binary number system. Unlike traditional algebra, Boolean algebra is easy to learn. It operates only on two numbers 0 & 1. There are only three operations: addition, multiplication, and complementation. The algebra of a number system basically describes how to perform arithmetic using the operators of the system acting upon the system's variables.

Boolean algebra describes the arithmetic of a two-state system and is therefore the mathematical language of digital electronics. The variables in Boolean algebra are represented by symbols such as A, B, C, X, Y etc. The rules of Boolean algebra are different from those of conventional algebra in the following aspects:

- Symbols used in Boolean algebra do not represent numerical values.
- Arithmetic operations (addition, subtraction, multiplication, division etc.) is not performed are not performed in Boolean algebra. Example: 1+1 is not 2.
- There are no fractions, negative numbers, squares etc.
- Boolean algebra allows only two possible values (0 and 1) or (L or H) for any variable.

These variables represent the input and output state (e.g. voltage in a circuit). These states can be represented by '0' or '1' in Boolean algebra.

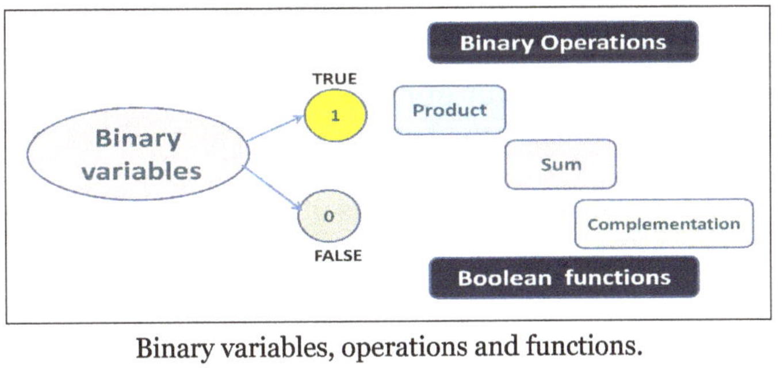

Binary variables, operations and functions.

Figure above indicates that in the Boolean algebra, Binary variables can take either of two values 0 and 1 (TRUE or FALSE), where as in traditional algebra variable can have any value up to infinity. There are three basic operations:

- Boolean product.
- Boolean Sum.
- Boolean complementation.

Boolean functions are implemented using combination of these operations. Addition and multiplication are two basic operations performed on the binary variables. Boolean algebra has only three operators: NOT, AND and OR. Boolean algebra provides techniques for describing, analyzing and designing their operation.

The variables can have only two values. Binary 1is used for HIGH and Binary 0 for LOW. Complement of a variable is represented by over bar (O) or ('). Ring of a variables is represented by

(+) sign between them. Logical ANDing of two or more variables is represented by a dot between them. Sometimes the dot between the variables may be ignored. The operation of the three basic gates and their Boolean descriptions, are Y = A’ or¯, Y = A.B and Y = A + B.

Boolean function has at least a variable and a operator. A variable is usually an uppercase letter to represent a logical quantity. A variable can have a 0 or 1 value. Complement is the inverse of a variable and indicated by a bar or single quote. E.g. complement of A is known as Not A or A bar. Boolean expression can contain sum or product terms or combination of both. Boolean expression can contain sum or product terms or combination of both.

$$Y = A\ .B$$

$$Z = C.\ D.\ E$$

$$W = F.G.H.K$$

As indicated here Boolean expressions are in AND form i.e. containing only product terms. Whereas P=A+B and Q=C+D+E indicates Boolean expressions in sum form. We can even get sum and product terms in an expression as shown below:

$$F(A,B,C,D) = (A.B) + (C.D)$$

Postulates, Laws and Theorems of Boolean Algebra

Boolean algebra is useful mathematical tool normally used to analyze a logic circuit and express its operation mathematically. It also plays an important role in designing digital system. The basic laws and theorems are normally utilized for minimization of Boolean expressions. Let us first discuss the important postulates of Boolean algebra.

Tables below show some of the important postulates of Boolean algebra. Axioms or Postulates are a set of basic logical expressions those are accepted as it is without proof. These postulates are normally used to build a set of useful laws and provide foundation to Boolean algebra. Really speaking postulates are nothing but representation of basic logical operations with binary values. Each postulate can be explained with the help of basic operations performed by AND, OR and NOT gates. Boolean algebra has its own set of fundamental rules which differ from the real algebra.

Table: Postulates of Boolean algebra.

OR Form		AND Form	
Real Algebra	Boolean Algebra	Real Algebra	Boolean Algebra
0+0 = 0	0+0 = 0	0.0 = 0	0.0 = 0
0+1 = 1	0+1 = 1	0.1 = 0	0.1 = 0
1+0 = 1	1+0 = 1	1.0 = 0	1.0 = 0
1+1 = 2	1+1 = 1	1.1 = 1	1.1 = 1

Similarly, 0’ = 1 and 1’ = 0 are two additional postulates related to complementation i.e. inverter or NOT gate. Many theorems of Boolean algebra are based on these postulates, which can be used to simplify Boolean expressions. Let us now discuss Boolean theorems. Let us now understand simple rules of Boolean algebra and study them in comparison with the real algebra.

Laws of Boolean Algebra

The rules, laws and theorems of Boolean algebra can be used to simplify many a complex Boolean expression and also to transform the given expression into a more useful and meaningful equivalent expression. Similar to real algebra, the Boolean algebra also possess certain well-defined rules and laws. Let us now discuss the most important laws of Boolean algebra.

- Identity Law: A term ANDed with a '1' or ORed with '0' equal to that term.
    - A.1=A [A variable ANDed with 1 is always equal to the variable].
    - A+0=A [A variable ORed with 0 is equal to the variable].

Table: Verification of Identity law.

A	LHS=A.1	RHS
0	0.1=0	0
1	1.1=1	1

A	LHS=A+0	RHS
0	0+0=0	0
1	1+0=1	1

- Null Law: A term ANDed with a '0' equals to 0 or ORed with '1' will equal 1.
    - A.0=0 [A variable ANDed with 0 is always equal to 0].
    - A+1=A [A variable ORed with 1 is always equal to 1].

Table: Verification of Null law.

A	LHS=A.0	RHS=0
0	0.0=0	0
1	1.0=0	0

A	LHS=A+1	RHS=1
0	0+1=1	1
1	1+1=1	1

- Idempotent Law: An input that is ANDed or ORed with itself is equal to the input.
    - A.A = A [A variable ANDed with itself is always equal to the variable].
    - A+A =A [A variable ORed with itself is always equal to the variable].

Table: Verification of Idempotent law.

A	LHS=A+A	RHS=A
0	0+0=0	0
1	1+1=1	1

A	LHS=A.A	RHS=0
0	0.0=0	0
1	1.1=1	1

- Inverse Law: A term ANDed with its complement equals '0' and a term ORed with its complement equal 1.
    - A. =0 [A variable ANDed with its complement is always equal to 0].
    - A+ =1 [A variable ORed with its complement is always equal to 1].

Table: Verification of Inverse law.

A	LHS=A.A'	RHS=A
0	0.1=0	0
1	1.0=1	0

X	LHS=A+A'	RHS=1
0	0+1=1	1
1	1+0=1	1

- Double complementation: A term that is complemented twice is equal to the original term.
  - $\bar{\bar{A}}$ = A [A double complementation of a variable is always equal to the variable].
- Commutative Law: This law indicates that the order of application of two separate terms is not important.
  - A.B = B.A [The order in which two variables are ANDed makes no difference on the Result].
  - A+B=B+A [The order in which two variables are ORed makes no difference on the Result].

A	B	A.B	B.A
0	0	0.0=0	0.0=0
0	1	0.1=0	1.0=0
1	0	1.0=0	0.1=0
1	1	1.1=1	1.1=1

A	B	A+B	B+A
0	0	0+0=0	0+0=0
0	1	0+1=1	1+0=1
1	0	1+0=1	0+1=1
1	1	1+1=1	1+1=1

Verification of Commutative law:

- Associative Law: This law enables the removal of the brackets from the expression and regrouping of the variables.
  - A.(B.C) = (A.B).C
  - A+(B+C) = (A+B) + C
- Distributive Law: This law allows the multiplying or factoring out an expression:
  - A.(B+C) = A.B + A.C
  - A+(B.C) = (A+B).(B+C)
- Absorptive Law: This law permits a reduction of a complicated expression to a simpler one by absorbing similar terms.
  - A.(A+B) = A
  - A+(A.B) = A

Simplification of Boolean Expression

Let us consider several examples to understand the applications of Boolean algebra in simplifying the expressions and to reduce the number of variable or the terms present in the Boolean function or expression.

Example:

$$\begin{aligned} A+(A.B) &= A.A+A.B \text{ (distributive law)} \\ &= A+AB \\ &= A.(1+B) \\ &= A.1 = A \end{aligned}$$

Example: Simplify the Boolean Expression,

$$\begin{aligned} f(X,Y,Z) &= \overline{X}.\overline{Y}+\overline{X}.\overline{Z}+Y.Z+X.\overline{Z} \\ &= \overline{Z}(X+\overline{X})+\overline{X}.\overline{Y}+Y.Z \\ &= \overline{X}.\overline{Y}+Y+Z \\ &= \overline{X}+Y+\overline{Z} \end{aligned}$$

Example: Simplify the Boolean Expression,

$$\begin{aligned} f(A,B,C) &= (A+\overline{B}.C)+(A+\overline{B}.C).M \\ &= (A+\overline{B}.C).(1+M) \\ &= (A+\overline{B}.C).1 \\ &= (A+\overline{B}.C) \end{aligned}$$

De Morgans's Theorems

De Morgan, a mathematician and a logician, proposed two theorems to enrich the Boolean algebra.

Theorem: The complement of product is equal to the sum of complements. It also states that the complement of two or more variables ANDed together is equivalent to ORing of complements of individual two or more variables.

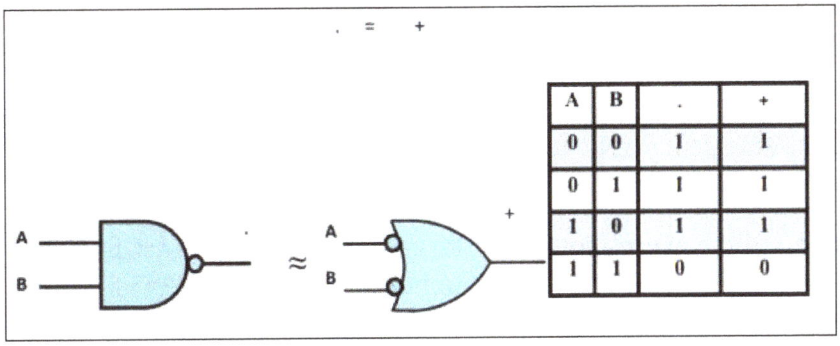

From the equivalent circuits and truth table, it is evident the NAND gate is equivalent to Bubbled OR gate. This law is applicable and extended to any number of variables and combination of variables. For example, ... = + + + +

Theorem: The complement of sum is equal to the product of complements. It also states that the complement of two or more variables ORed together is equivalent to the ANDing of the complements of individual two or more variables.

+ = .

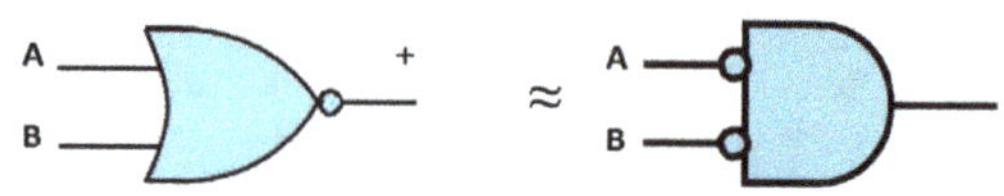

A	B	+	.
0	0	1	1
0	1	0	0
1	0	0	0
1	1	0	0

Verification of De Morgan's previous theorem.

From the equivalent circuits and truth table, it is evident the NOR gate is equivalent to Bubbled AND gate. This law is applicable and extended to any number of variables and combination of variables. For example, + + + + = ...

At this stage it is important to note that De Morgan's theorems can be used for transformation of Boolean expressions from sum-of-complements or complement-of-sum forms and vice a versa.

Example: Apply De Morgan's theorem,

$$\overline{\overline{(A+B)}+C} = \overline{\overline{(A+B)}}.\overline{C}$$
$$= (A+B).\overline{C} = A.\overline{C} + B\overline{C}$$

Points to Remember Regarding De Morgan's

- De Morgan's theorem allows large bars in a Boolean Expression to be broken up into smaller bars over individual variables.
- De Morgan's theorem says that a large bar over several variables can be broken between the variables if the sign between the variables is changed.
- De Morgan's theorem can be used to prove that a NAND gate is equal to an OR gate with inverted inputs.
- De Morgan's theorem can be used to prove that a NOR gate is equal to an AND gate with inverted inputs.
- In order to reduce expressions with large bars, the bars must first be broken up. This means that in some cases, the first step in reducing an expression is to use De Morgan's theorem.

Visualization of Signal Flow using De Morgan's Theorem

In many complex logic circuits, to determine the output of the logic circuit one needs to prepare a truth table. Truth table for logic circuit with more than 3 variables requires more than 8 combinations at input. In order determine the active condition for the output it is possible to visualize the signal flow without referring to the truth table. Bubble pushing is a technique to apply De Morgan's theorem directly to the logic diagram.

- Inspect the circuit starting from output and go back towards input.
- Identify the active condition for the output.
- Internally change the logic gate NAND to its equivalent OR and NOR to its equivalent AND so as keep active levels same for the interconnections.
- Active HIGH an output is connected to active HIGH input or bubbled output be connected to bubbled input.

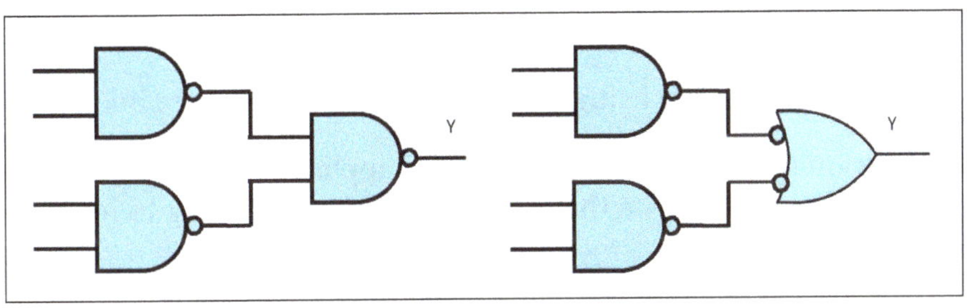

Visualization of signal flow.

Figure above shows a simple circuit using interconnected NAND circuit with 4 inputs A,B,C & D to produce output Y. One can use above mentioned procedure to visualize the single flow using De Morgan's theorem. Replacing output NAND by its De Morgan's equivalent i.e. bubbled OR gate produces interconnections of Bubbled output to bubble input gate. The output Y is HIGH if either output of first NAND or Second NAND is LOW. The NAND gate output is low only when both the inputs are HIGH. Thus to get Y=1, either A=B=1 or C=D=1. Please note that there is no need to prepare the truth table to determine the active state of the circuit.

Logic gates can be De Morganized so that bubbles appear on inputs or outputs in order to satisfy signal conditions rather than specific logic functions. An active-low signal should be connected to a bubble on the input of a logic gate.

Duality Principle

Axioms or postulates for Boolean algebra are always come in pairs. If you swap + with. and 0 and 1, in one of the pair you will get the second. The duals of a Boolean expression are obtained by swapping these operators and 0 and 1. Note that the dual of any theorem in Boolean algebra is also a theorem.

The principle of duality is an important property in Boolean algebra, specifically for proving the laws and theorems. It states that every algebraic expression deducible from the postulates of Boolean algebra remains valid if the operators and identity elements are interchanged. If the dual of an algebraic expression is desired, we simply interchange OR and AND operators and replace 1's by 0's and 0's by 1's.

For example: The dual of Boolean expression A.B + A.B is (A+B).(A+B). It is observed that the principle of duality is used extensively in proving the laws and theorems of Boolean algebra. As Boolean algebra deals with variables consisting of only two elements then we can prove both the sides are valid and equivalent.

Realization of Logic Circuit/Boolean Algebra

Logic circuits can be realized from the Boolean expression or Boolean function can be realized for given logic circuit. Let us understand the procedure for the same.

Realize a logic circuit from boolean expression:

- Identify the number of terms present in logical expressions.
- Determine the logic gates required to implement each term.
- Represent each term by a logic gate.
- Interconnect al the logic gates to realize the Boolean expression.
- Label the gates appropriately using the variables indicated by the Boolean expression.

Realize a logic expression from a logic circuit: In many applications, it is necessary to analyze the given logic circuit and we have to write the Boolean expression relating to input and output. The procedure is as follows:

- Begin with the input side.
- Write the expression for output of each gate.
- Repeat the process for all remaining gates and proceed towards the output.
- Combine all the terms to get complete Boolean expression for output.

Karnaugh Maps

In many digital circuits and practical problems we need to find expression with minimum variables. We can minimize Boolean expressions of 3, 4 variables very easily using K-map without using any Boolean algebra theorems. K-map can take two forms Sum of Product (SOP) and Product of Sum (POS) according to the need of problem. K-map is table like representation but it gives more information than TRUTH TABLE. We fill grid of K-map with 0's and 1's then solve it by making groups.

Steps to Solve Expression using K-Map

- Select K-map according to the number of variables.
- Identify minterms or maxterms as given in problem.

- For SOP put 1's in blocks of K-map respective to the minterms (0's elsewhere).
- For POS put 0's in blocks of K-map respective to the maxterms (1's elsewhere).
- Make rectangular groups containing total terms in power of two like 2,4,8.. (except 1) and try to cover as many elements as you can in one group.
- From the groups made in step 5 find the product terms and sum them up for SOP form.

It is a graphical method, which consists of 2n cells for 'n' variables. The adjacent cells are differed only in single bit position.

K-maps for 2 to 5 Variables

K-map method is most suitable for minimizing Boolean functions of 2 variables to 5 variables. Now, let us discuss about the K-Maps for 2 to 5 variables one by one.

2 Variable K-map

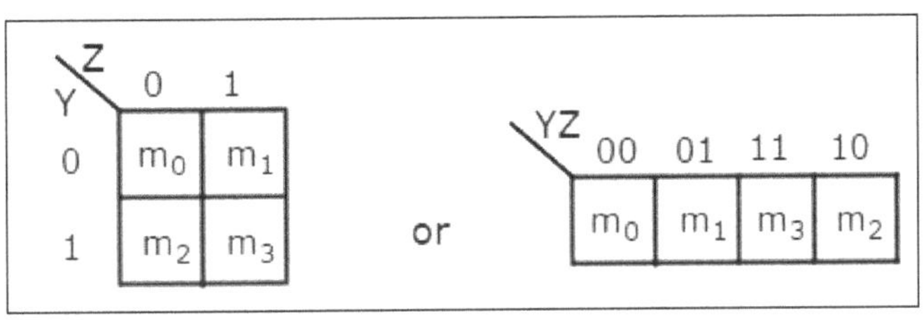

The number of cells in 2 variable K-map is four, since the number of variables is two. The above figure shows 2 variable K-Map:

- There is only one possibility of grouping 4 adjacent min terms.
- The possible combinations of grouping 2 adjacent min terms are {(m_0, m_1), (m_2, m_3), (m_0, m_2) and (m_1, m_3)}.

3 Variable K-map

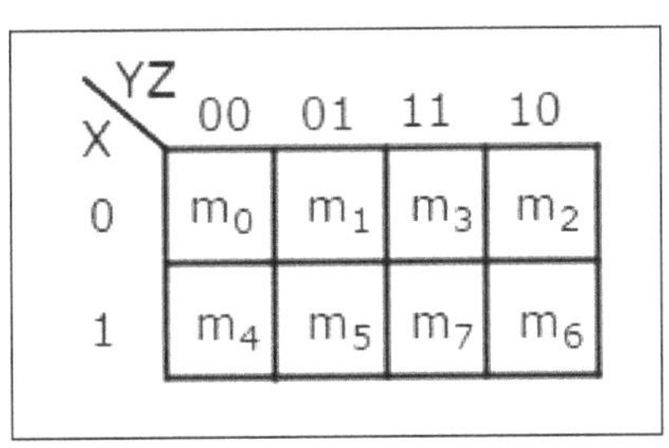

The number of cells in 3 variable K-map is eight, since the number of variables is three. The above figure shows 3 variable K-map:

- There is only one possibility of grouping 8 adjacent min terms.
- The possible combinations of grouping 4 adjacent min terms are {(m_0, m_1, m_3, m_2), (m_4, m_5, m_7, m_6), (m_0, m_1, m_4, m_5), (m_1, m_3, m_5, m_7), (m_3, m_2, m_7, m_6) and (m_2, m_0, m_6, m_4)}.

- The possible combinations of grouping 2 adjacent min terms are $\{(m_0, m_1), (m_1, m_3), (m_3, m_2), (m_2, m_0), (m_4, m_5), (m_5, m_7), (m_7, m_6), (m_6, m_4), (m_0, m_4), (m_1, m_5), (m_3, m_7)$ and $(m_2, m_6)\}$.
- If x=0, then 3 variable K-map becomes 2 variable K-map.

4 Variable K-map

WX \ YZ	00	01	11	10
00	m_0	m_1	m_3	m_2
01	m_4	m_5	m_7	m_6
11	m_{12}	m_{13}	m_{15}	m_{14}
10	m_8	m_9	m_{11}	m_{10}

The number of cells in 4 variable K-map is sixteen, since the number of variables is four. The above figure shows 4 variable K-Map.

- There is only one possibility of grouping 16 adjacent min terms.
- Let R_1, R_2, R_3 and R_4 represents the min terms of first row, second row, third row and fourth row respectively. Similarly, C_1, C_2, C_3 and C_4 represents the min terms of first column, second column, third column and fourth column respectively. The possible combinations of grouping 8 adjacent min terms are $\{(R_1, R_2), (R_2, R_3), (R_3, R_4), (R_4, R_1), (C_1, C_2), (C_2, C_3), (C_3, C_4), (C_4, C_1)\}$.
- If w=0, then 4 variable K-map becomes 3 variable K-map.

5 Variable K-map

V=0

WX \ YZ	00	01	11	10
00	m_0	m_1	m_3	m_2
01	m_4	m_5	m_7	m_6
11	m_{12}	m_{13}	m_{15}	m_{14}
10	m_8	m_9	m_{11}	m_{10}

V=1

WX \ YZ	00	01	11	10
00	m_{16}	m_{17}	m_{19}	m_{18}
01	m_{20}	m_{21}	m_{23}	m_{22}
11	m_{28}	m_{29}	m_{31}	m_{30}
10	m_{24}	m_{25}	m_{27}	m_{26}

The number of cells in 5 variable K-map is thirty-two, since the number of variables is 5. The above figure shows 5 variable K-map.

- There is only one possibility of grouping 32 adjacent min terms.
- There are two possibilities of grouping 16 adjacent min terms. i.e., grouping of min terms from m_0 to m_{15} and m_{16} to m_{31}.
- If v=0, then 5 variable K-map becomes 4 variable K-map.

In the above all K-maps, we used exclusively the min terms notation. Similarly, you can use exclusively the Max terms notation.

Minimization of Boolean Functions using K-maps

If we consider the combination of inputs for which the Boolean function is '1', then we will get the Boolean function, which is in standard sum of products form after simplifying the K-map. Similarly, if we consider the combination of inputs for which the Boolean function is '0', then we will get the Boolean function, which is in standard product of sums form after simplifying the K-map. Follow these rules for simplifying K-maps in order to get standard sum of products form:

- Select the respective K-map based on the number of variables present in the Boolean function.
- If the Boolean function is given as sum of min terms form, then place the ones at respective min term cells in the K-map. If the Boolean function is given as sum of products form, then place the ones in all possible cells of K-map for which the given product terms are valid.
- Check for the possibilities of grouping maximum number of adjacent ones. It should be powers of two. Start from highest power of two and upto least power of two. Highest power is equal to the number of variables considered in K-map and least power is zero.
- Each grouping will give either a literal or one product term. It is known as prime implicant. The prime implicant is said to be essential prime implicant, if atleast single '1' is not covered with any other groupings but only that grouping covers.
- Note down all the prime implicants and essential prime implicants. The simplified Boolean function contains all essential prime implicants and only the required prime implicants.

If outputs are not defined for some combination of inputs, then those output values will be represented with don't care symbol 'x'. That means, we can consider them as either '0' or '1'. If don't care terms also present, then place don't cares 'x' in the respective cells of K-map. Consider only the don't cares 'x' that are helpful for grouping maximum number of adjacent ones. In those cases, treat the don't care value as '1'.

Example: Let us simplify the following Boolean function, f W, X, Y, Z = WX'Y' + WY + W'YZ' using K-map. The given Boolean function is in sum of products form. It is having 4 variables W, X, Y & Z. So, we require 4 variable K-map. The 4 variable K-map with ones corresponding to the given product terms is shown in the following figure:

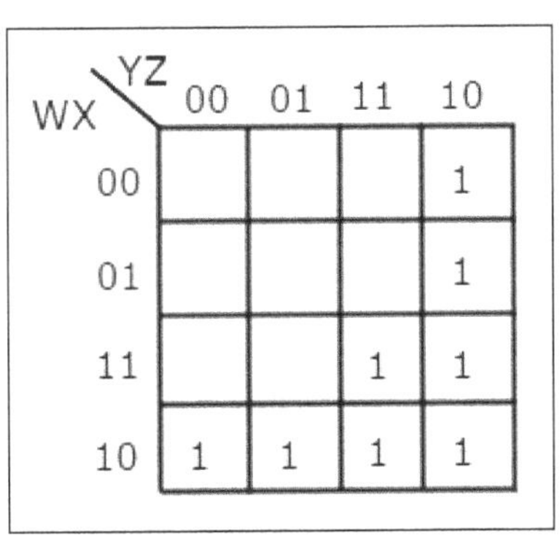

Here, 1s are placed in the following cells of K-map:

- The cells, which are common to the intersection of Row 4 and columns 1 & 2 are corresponding to the product term, WX'Y'.
- The cells, which are common to the intersection of Rows 3 & 4 and columns 3 & 4 are corresponding to the product term, WY.
- The cells, which are common to the intersection of Rows 1 & 2 and column 4 are corresponding to the product term, W'YZ'.

There are no possibilities of grouping either 16 adjacent ones or 8 adjacent ones. There are three possibilities of grouping 4 adjacent ones. After these three groupings, there is no single one left as ungrouped. So, we no need to check for grouping of 2 adjacent ones. The 4 variable K-map with these three groupings is shown in the following figure:

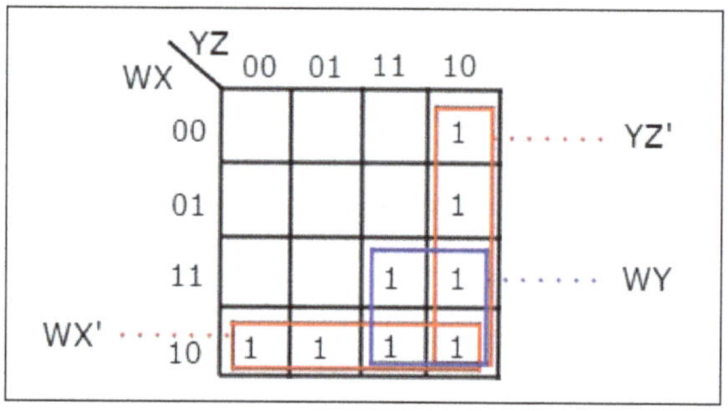

Here, we got three prime implicants WX', WY & YZ'. All these prime implicants are essential because of following reasons:

- Two ones (m_8 & m_9) of fourth row grouping are not covered by any other groupings. Only fourth row grouping covers those two ones.
- Single one (m_{15}) of square shape grouping is not covered by any other groupings. Only the square shape grouping covers that one.
- Two ones (m_2 & m_6) of fourth column grouping are not covered by any other groupings. Only fourth column grouping covers those two ones.

Therefore, the simplified Boolean function is,

$$f = WX' + WY + YZ'$$

Follow these rules for simplifying K-maps in order to get standard product of sums form:

- Select the respective K-map based on the number of variables present in the Boolean function.
- If the Boolean function is given as product of Max terms form, then place the zeroes at respective Max term cells in the K-map. If the Boolean function is given as product of sums form, then place the zeroes in all possible cells of K-map for which the given sum terms are valid.

- Check for the possibilities of grouping maximum number of adjacent zeroes. It should be powers of two. Start from highest power of two and upto least power of two. Highest power is equal to the number of variables considered in K-map and least power is zero.
- Each grouping will give either a literal or one sum term. It is known as prime implicant. The prime implicant is said to be essential prime implicant, if atleast single '0' is not covered with any other groupings but only that grouping covers.
- Note down all the prime implicants and essential prime implicants. The simplified Boolean function contains all essential prime implicants and only the required prime implicants.
- If don't care terms also present, then place don't cares 'x' in the respective cells of K-map. Consider only the don't cares 'x' that are helpful for grouping maximum number of adjacent zeroes. In those cases, treat the don't care value as '0'.

Example: Let us simplify the following Boolean function, $f(X,Y,Z)=\prod M(0,1,2,4)$ using K-map. The given Boolean function is in product of Max terms form. It is having 3 variables X, Y & Z. So, we require 3 variable K-map. The given Max terms are M_0, M_1, M_2 & M_4. The 3 variable K-map with zeroes corresponding to the given Max terms is shown in the following figure:

X \ YZ	00	01	11	10
0	0	0		0
1	0			

There are no possibilities of grouping either 8 adjacent zeroes or 4 adjacent zeroes. There are three possibilities of grouping 2 adjacent zeroes. After these three groupings, there is no single zero left as ungrouped. The 3 variable K-map with these three groupings is shown in the following figure:

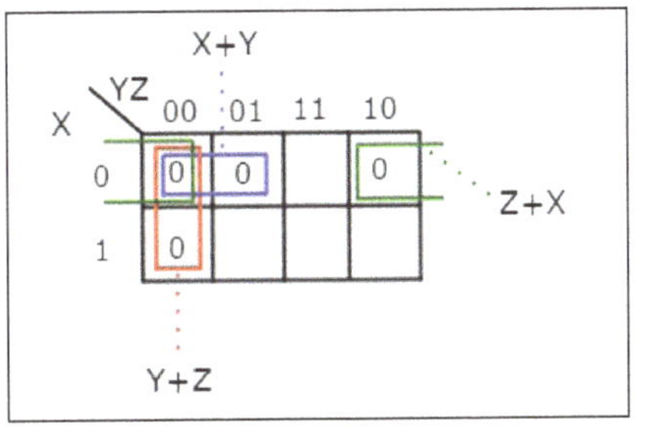

Here, we got three prime implicants X + Y, Y + Z & Z + X. All these prime implicants are essential because one zero in each grouping is not covered by any other groupings except with their individual groupings. Therefore, the simplified Boolean function is,

$$f = X+Y \ . \ Y+Z \ . \ Z+X$$

In this way, we can easily simplify the Boolean functions up to 5 variables using K-map method. For more than 5 variables, it is difficult to simplify the functions using K-maps. Because, the number of cells in K-map gets doubled by including a new variable.

Encoder

Numbers are usually coded in one form or another so as to represent or use it as required. In modern communication system, data communication involves a process encoding and decoding. Encoding is the process of converting human readable characters into binary information. Decoding is the reverse process of conversion of binary or encoded format into human readable code. This process is also referred to as code conversion.

There are numerous encoding and decoding methods exist. The oldest code of all, originally employed in the landline communication for telegraph application was the Morse code. The code used by almost all computers is the ASCII (American Standard Code for Information Interchange) code. ASCII utilizes uppercase, lowercase alphabetic characters, numeral, punctuation marks and common symbols. In addition, digital system designers use Binary, BCD, Octal Hexadecimal, Decimal, Gray, Excess-3 code and several other codes. In data communication, a special form of encoding is used to convert binary digits too high to low or low to high transitions.

Encoding and decoding are also used in context with the process of analog to digital conversion or digital to analog conversion. In this sense, code conversion term can apply to any form of data including text, images, audio, video, multimedia, computer codes, sensor signals. Encoding should not be confused with encryption.

Code Conversion

Modern digital system like calculators, mobile phones, fax machines, computers and other digital gadgets use binary numbers 1 and 0 for processing and storage. However, our language of communication is in the form of decimal numbers and alphabetic characters. It is difficult for human being to understand information represented in binary. Therefore, there is a need of a group of devices to convert human readable language to machine or binary language. To achieve this task, several binary codes are developed. The process of generating binary codes is known as encoding and the reverse process is known as decoding. Let us consider a simple example to demonstrate the role of encoder and decoder. Figure below indicates simple block diagram of generalized digital system.

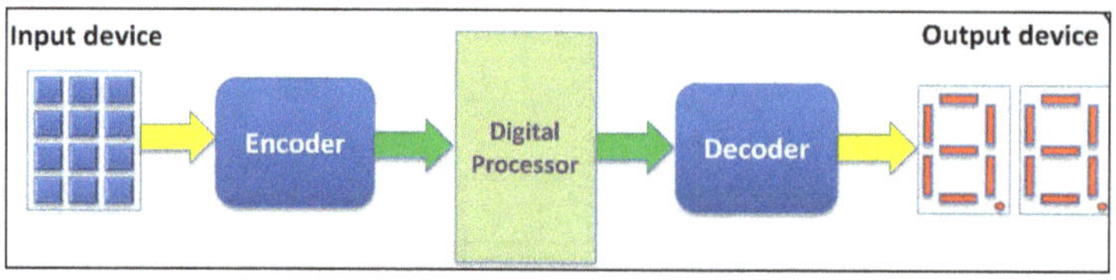

Simple block diagram of Digital system.

A digital system cannot be complete without encoders and decoders. It contains five basic building blocks. Keyboard and seven segment display are the input and output devices respectively. The encoders essentially receive the input from the keyboard. Many keyboards use encoders to produce a machine recognizable code i.e. binary and decoder accepts the processed binary code and converts back in the format useful for output devices like seven segment display, CRT, Printers etc. BCD, binary, Octal, Hexadecimal, ASCII and decimal are the most common codes used in the digital systems. Thus encoders and decoders play an important role in code conversion process in digital system. One can make comparison of the above system with the functional blocks of hand held calculator.

The process of converting from human readable code to machine readable code i.e. binary is known as encoding. An encoder is a combinational circuit that coverts more familiar numbers, symbols or character into binary code. An encoder has a number of input lines but only one of them is activated at a time representing a digit or character and produces a binary code depending on which input is activated. Figure below is the logic symbol of encoder with 'm' inputs and 'n' outputs. In short, it is multiple inputs and multiple outputs device with proper conversion system. Note that encoder performs the reverse operation of the decoder.

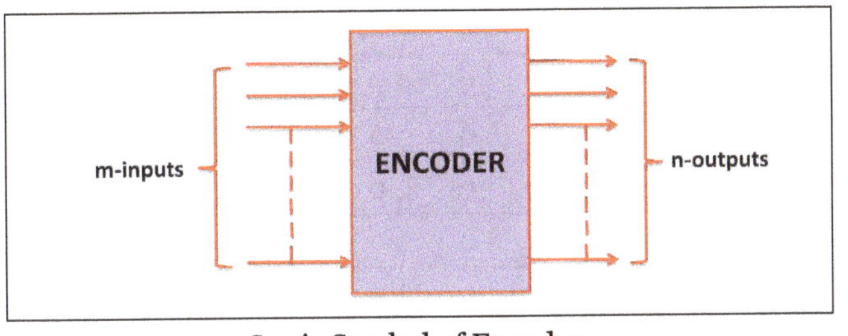

Logic Symbol of Encoder.

An encoder has 'm' number of input lines and 'n' number of output lines. The numbers of outputs (n) are always less than number of inputs (m). Here the number of inputs can have following relation:

$$m \le 2^n$$

Thus, the encoder accepts an m-bit input digital word and converts it into an n-bit another digital word. Encoders can also be devised to encode various symbols and characters for the keypads and keyboards. Some of the most commonly used encoders are – (1) Linear encoders are octal to binary, Decimal to BCD and Hexadecimal to binary where normal encoding is implemented and (2) Priority encoders.

Liner Encoders

Linear/normal encoders accepts m inputs, only one is activated at a given time and produces n-bit output code. As the numbers of output lines are 'n', it is possible to encode maximum up to 2^n inputs. In many digital systems, octal, decimal or hexadecimal numbers are used as inputs.

Octal to Binary Encoder

An octal to binary encoder is also known as 8-line to 3-line encoder. It accepts 8-inputs and produces a 3-bit output corresponding to the activated octal input. Figure below shows the logic symbol of Octal to binary encoder.

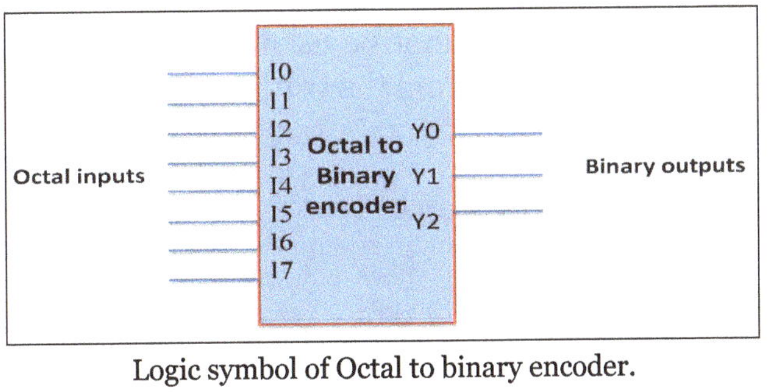

Logic symbol of Octal to binary encoder.

The truth table of octal to binary encoder is shown in Table. There are eight inputs corresponding 8 octal inputs (I0, I1, I2, I3, I4, I5, I6, I7) and 3 Outputs (Y0, Y1, Y2). Let us look at the table carefully. Note that in encoders only one input is activated (logic 1) and other inputs are not activated (i.e. at logic 0).

Table: Truth – Table for Octal Binary Encoder.

Octal Inputs								Binary Outputs		
I0	I1	I2	I3	I4	I5	I6	I7	Y2	Y1	Y0
1	0	0	0	0	0	0	0	0	0	0
0	1	0	0	0	0	0	0	0	0	1
0	0	1	0	0	0	0	0	0	1	0
0	0	0	1	0	0	0	0	0	1	1
0	0	0	0	1	0	0	0	1	0	0
0	0	0	0	0	1	0	0	1	0	1
0	0	0	0	0	0	1	0	1	1	0
0	0	0	0	0	0	0	1	1	1	1

In the truth table, the input variable I0 represents the least significant digit (LSD) and and I7 represents most significant digit (MSD). Similarly, in the outputs Y0 represents least significant bit (LSB) and Y2 is the most significant bit (MSB). The truth table includes only all valid combinations of the inputs. The valid combinations are those which have exactly one input equal to logic 1 while all other inputs are logic 0's.

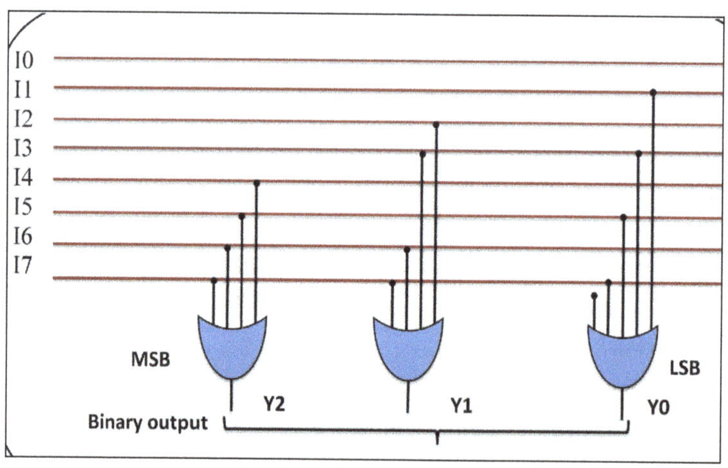

Octal to Binary Encoder.

As the number of inputs are 8, K-maps cannot be used to derive the Boolean expressions. The Boolean expression can be directly derived from the truth table by visual inspection. Let us obtain the Boolean expression for each output.

Output Y0 is 1 if any of the inputs I1 or I3 or I5 or I7 is 1. Then, the Boolean expression,

$Y0 = I1 + I3 + I5 + I7$.

Similarly, $Y1 = I2 + I3 + I6 + I7$ and $Y2 = I4 + I5 + I6 + I7$.

From these Boolean expressions, the Octal to Binary Encoder can be implemented by using simply three 4 input Or gates. Figure indicates the logic diagram for Octal to Binary Encoder.

This octal to binary encoder design has few drawbacks. In the encoder, at a time only one input must be active. (1) If number of active inputs is greater than one then the output is incorrect. (2) Similarly, when no inputs are active then all outputs are low which is also similar to the output generated by active input I0. To resolve first problem, one must design and use priority encoder where the output corresponding to input with higher weightage is generated. The solution to the second drawback is to generate additional output signal to check the validity of inputs. If no inputs are active, data is invalid.

Decimal to BCD Encoder

A decimal to BCD (binary coded decimal) encoder is also known as 10-line to 4-line encoder. It accepts 10- inputs and produces a4-bit output corresponding to the activated decimal input. Figure below shows the logic symbol of decimal to BCD encoder.

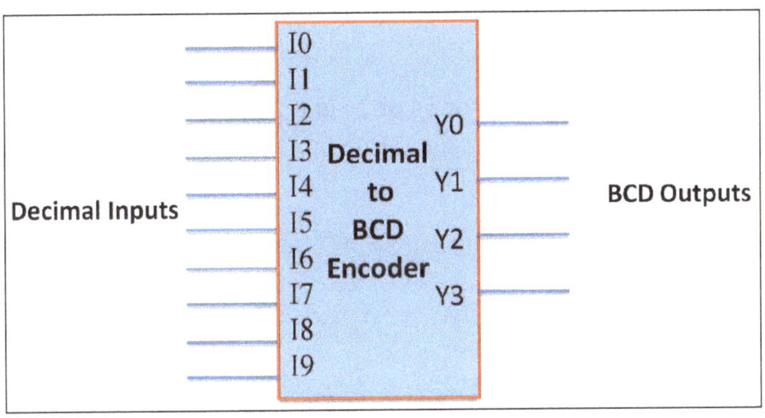

The truth table of Decimal to BCD encoder is shown in Table. There are ten inputs corresponding 10 decimal inputs (I0, I1, I2, I3, I4, I5, I6, I7, I8, I9) and 4 Outputs (Y0, Y1, Y2, Y3). Let us look at the table carefully. Note the encoder assumption that only one of the inputs is activated (logic 1) and other inputs are not activated (i.e. at logic 0).

Table: Truth – Table for Decimal to BCD Encoder.

Decimal Inputs										BCD Outputs			
I0	I1	I2	I3	I4	I5	I6	I7	I8	I9	Y3	Y2	Y1	Y0
1	0	0	0	0	0	0	0	0	0	0	0	0	0
0	1	0	0	0	0	0	0	0	0	0	0	0	1
0	0	1	0	0	0	0	0	0	0	0	0	1	0
0	0	0	1	0	0	0	0	0	0	0	0	1	1
0	0	0	0	1	0	0	0	0	0	0	1	0	0

0	0	0	0	0	1	0	0	0	0	0	1	0	1
0	0	0	0	0	0	1	0	0	0	0	1	1	0
0	0	0	0	0	0	0	1	0	0	0	1	1	1
0	0	0	0	0	0	0	0	1	0	1	0	0	0
0	0	0	0	0	0	0	0	0	1	1	0	0	1

In the truth table, the input variable I0 represents the least significant digit (LSD) and and I9 represents most significant digit (MSD). Similarly, in the outputs Y0 represents least significant bit (LSB) and Y3 is the most significant bit (MSB). The truth table includes only all valid combinations of the inputs. The valid combinations are those which have exactly one input equal to logic 1 while all other inputs are logic 0's.

As the number of inputs is 10, K-maps cannot be used to derive the Boolean expressions. The Boolean expression can be directly derived from the truth table by visual inspection. Let us obtain the Boolean expression for each output.

Output Y0 is 1 if any of the inputs I1 or I3 or I5 or I7 is 1. Then, the Boolean expression,

$$Y0 = I1 + I3 + I5 + I7 + I9.$$

Similarly,

$$Y1 = I2 + I3 + I6 + I7$$

$$Y2 = I4 + I5 + I6 + I7$$

$$Y3 = I8 + I9$$

From these Boolean expressions, the Decimal to BCD Encoder can be implemented by using simply three 4 OR gates. Figure indicates the logic diagram for Decimal to BCD Encoder. This decimal to BCD encoder design also provides few drawbacks as mentioned in Octal to Binary encoder. The solution to this problem is to use priority encoder and use of additional valid output in the encoder design.

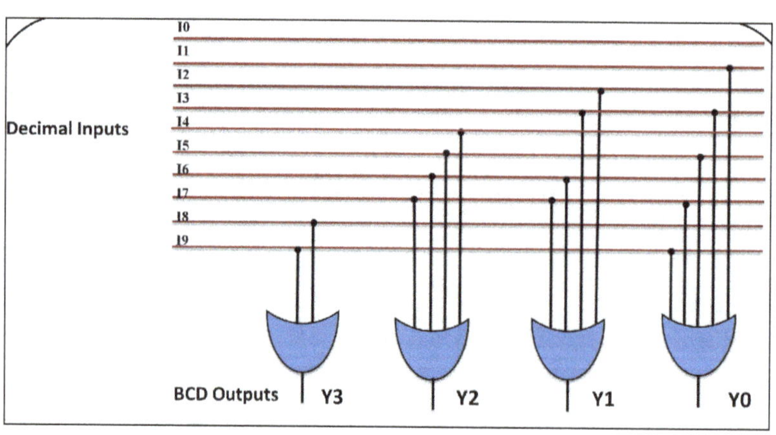

Decimal to BCD Encoder.

Hexadecimal to Binary Encoder

It accepts sixteen inputs (0, 1, 2, ...9, A,B,C,D,E,F) and produces 4-bit binary code corresponding to the activated Hexadecimal input. The logic diagram for Hexadecimal to Binary encoder can be developed on the similar line of Decimal to BCD encoder. The truth table for the same encoder can be prepared by simply writing four bit binary code for each activated hexadecimal input as shown in table.

Table: Truth table for Hexadecimal to Binary encoder.

Hexadecimal Inputs	Binary Outputs			
	Y3	Y2	Y1	Y0
0	0	0	0	0
1	0	0	0	1
2	0	0	1	0
3	0	0	1	1
4	0	1	0	0
5	0	1	0	1
6	0	1	1	0
7	0	1	1	1
8	1	0	0	0
9	1	0	0	1
A	1	0	1	0
B	1	0	1	1
C	1	1	0	0
D	1	1	0	1
E	1	1	1	0
F	1	1	1	1

Priority Encoder

Standard encoders generate the wrong output code when there is more than one input present at logic 1. This is one of the major disadvantages of the linear encoders. One of the simplest ways to overcome this problem is to prioritize the inputs. If there are more than one input at logic '1' then the actual code correspond to the highest designated priority are produced and all other inputs with lower priority will be ignored. This type of encoder is known as priority encoder.

4 to 2 Priority Encoder

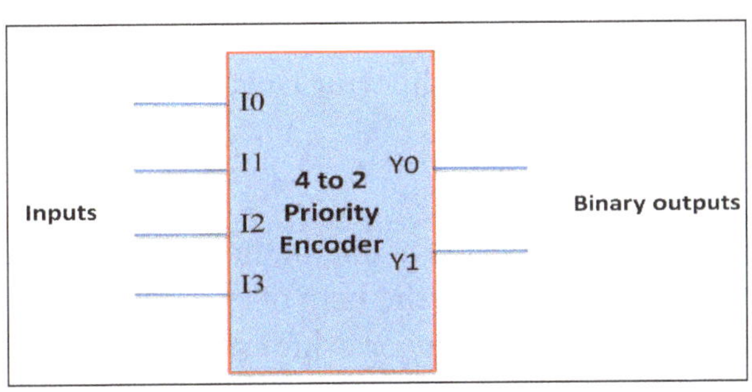

Figure above shows the logic symbol of 4 to 2 priority encoder. Priorities are given to the input lines. In this case input I3 is having highest priority than other inputs, I2 has second highest priority as compare to I1 and I0 and so on. If two or more inputs are at logic '1' at the same time, the input line with highest priority will be considered for generating binary output.

Table: Truth table of 4 to 2 priority encoder.

Inputs				Outputs	
I3	I2	I1	I0	Y1	Y0
0	0	0	0	X	X
0	0	0	1	0	0
0	0	1	X	0	1
0	1	X	X	1	0
1	X	X	X	1	1

There are four inputs I0 through I3 and two outputs Y0 and Y1. With these four inputs, the actual truth table would have been with 16 possible input conditions. But for simplicity a truth table is prepared wisely and contain only few conditions with the help of X – do not care condition. As shown in table. Out of the four inputs I3 has highest priority and I0 has lowest priority. That means when I3=1 then Y1 Y0 = 11 irrespective of the other inputs.

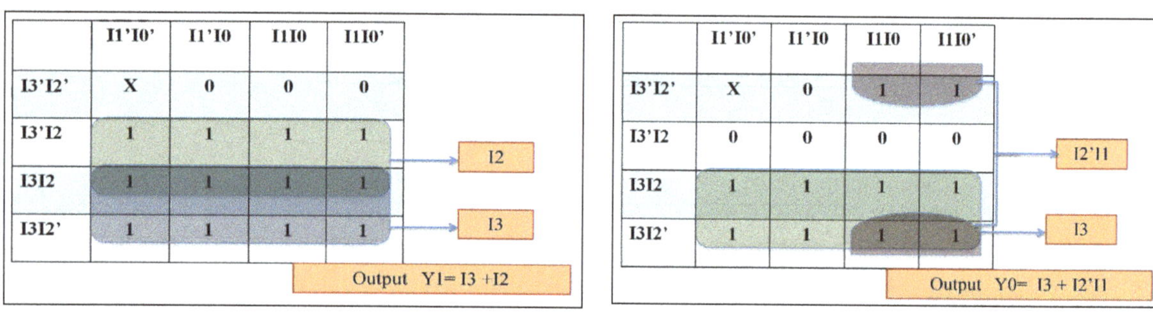

K-maps for the outputs of 4 to 2 line encoder.

The k-maps for two outputs Y1 and Y0 are shown in figure. The logic '1' is mapped for each output for all input conditions. For output Y1, it is possible to group 8 neighboring cells to form the octet. It is possible to form two octets. From these octet, we can eliminate three variables each. The remaining unchanged variables are written in the simplified for. Thus,

$$\text{Output } Y1 = I3 + I2$$

Similarly we can obtain Boolean expression for output Y0 using another K-map. After mapping the '1's in the map, it is possible to group neighboring cells to form Octet and Quad. Thus, we get,

$$\text{Output } Y0 = I3 + I2'I1$$

Using these Boolean expressions, it is possible to obtain the logic diagram for 4 to 2 priority encoder. The two binary outputs are obtained with the help of two OR gates, one AND gate and a NOT gate. The logic diagram for 4 to 2 priority encoder is indicated in figure.

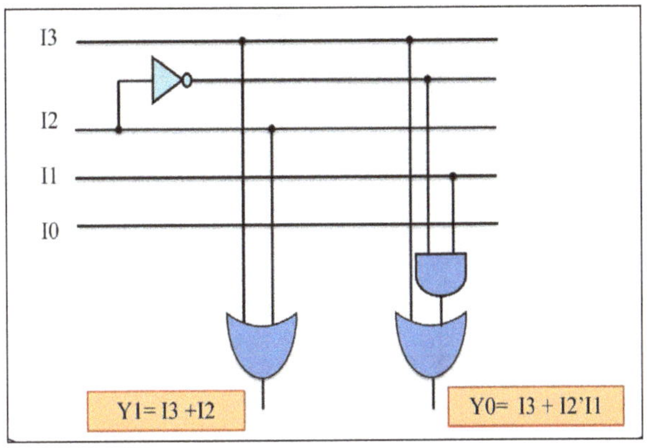

Logic diagram for 4 to 2 Priority Indicator.

If more than one input is active, the higher order input has priority over the lower order input. The higher value is encoded at the output. Further, one can introduce one more output for the priority encoder to indicate whether the output is valid or not. This output is also called as validity indicator. It is generated by simply ORing all possible inputs. If all the inputs are '0', the output is invalid. The output is valid (logic-1) only when at least one input is active.

Priority Encoders are also available in the Integrated Circuit form. The TTL priority encoders are:

- IC 74148 – Octal to Binary priority encoder.
- IC 74147 – Decimal to BCD priority encoder.

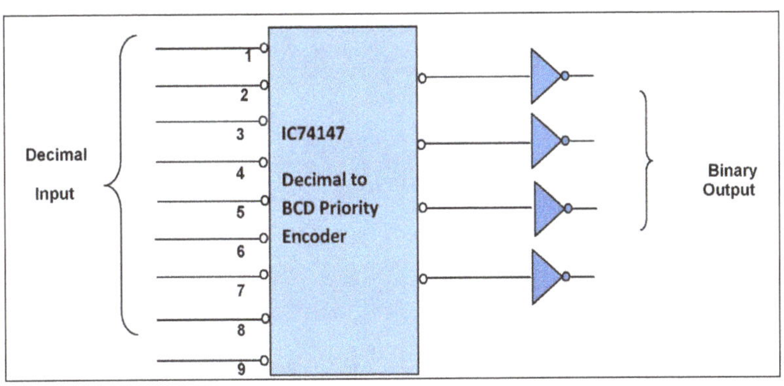

Logic Symbol of Decimal to BCD Priority encoder IC 74147.

Keyboard Encoder

Keyboards are generally used for data entry into the computer and digital systems. The Keyboards are either hexadecimal or ASCII type. For electronic calculators there are 10 switches (Keys) for decimal number 0 to 9 and some function keys. When a digit key is pressed the internal circuit understands which key is pressed and Keyboard encode will give a code corresponding to key pressed. A typical Keyboard encoder using IC 74147 a priority Encoder is shown in below figure.

There are 10 Push button switches with pull-up resistors connected to the positive Supply. The pull up resistor makes the input line high when the switch is not closed. When a switch is closed the corresponding line will be connected to ground. This provides low level to the corresponding input to the encoder. When key corresponding to decimal digit say 3 is pressed, the encoder IC generates a BCD code of 3 in complemented form. The generated binary code can be converted to proper BCD by using inverter at the output of the encoder. In this encoder, there are nine inputs with highest priority to number 9 and least to number 1.

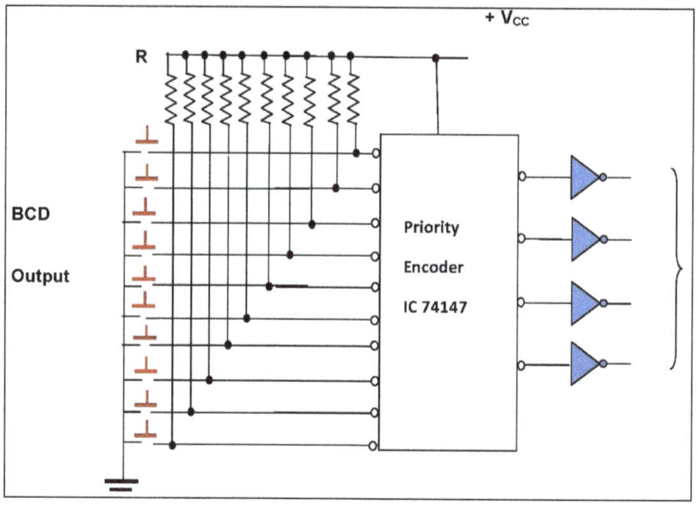

Applications of Encoder

Encoders are normally preferred where there is a need to convert information of code from one format to another. In addition to linear and priority encoders, encoders have many applications as listed below:

- Keyboard encoder for computers.
- Optical encoders –linear or rotary.
- Interfacing peripherals to microprocessors.
- Audio/video coding and transmission.

Decoder

Decoders are digital systems which are used for decoding the information. In other words, the decoders decrypt or retrieve the actual data from the received code. It converts the binary input code into the desired output format. In all Modern digital system like calculators, mobile phones, fax machines, computers and other digital gadgets, encoder is used at input side for encoding and decoder is used at output side for decoding. There is a wide variety of applications like data de-multiplexing, memory address decoding, seven segment display decoding, binary function generator etc.

Figure indicates the position of encoder and decoder in the digital system. Encoders receive the input from the keyboard to generate binary for further processing. Decoder accepts the processed binary code and converts back in the format useful for output devices like seven segment display, CRT, Printers etc.

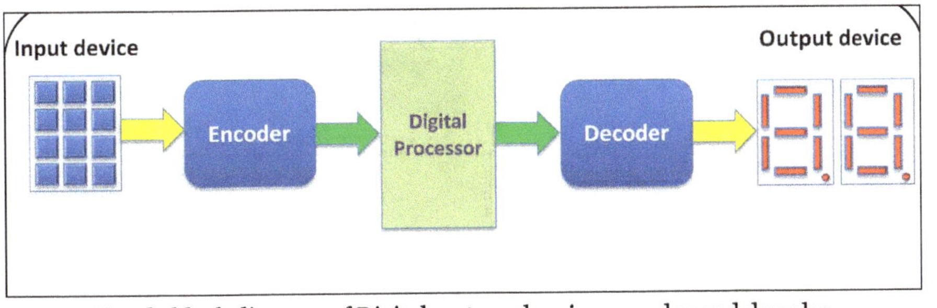

Simple block diagram of Digital system showing encoder and decoder.

Decoder is a combinational logic circuit that converts a binary code into the desired output signals. It is called decoder because it performs the reverse process of encoder. The process of converting binary input code into desirable output is known as decoding. Figure below is the logic symbol of decoder with 'n' inputs and 'm' outputs. In short, it is multiple inputs and multiple outputs device with proper conversion system. Note that decoder performs the reverse operation of the encoder.

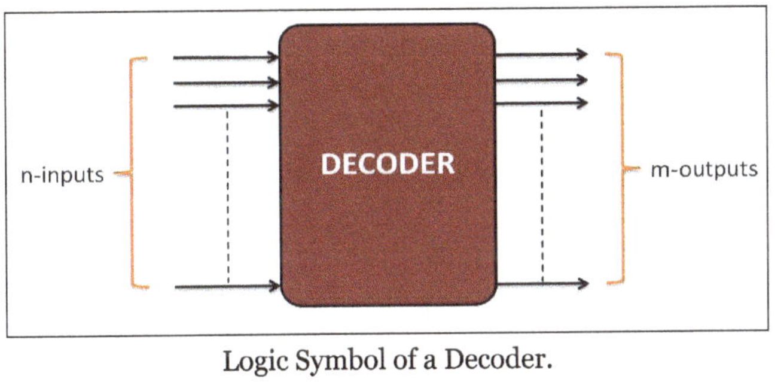

Logic Symbol of a Decoder.

A decoder has 'n' number of input lines and 'm' number of output lines. There are 2^n possible input combinations and the number of outputs (m) are less than or equal to 2^n. Here the number of outputs can have following relation:

$$m \geq 2^n$$

Decoder converts an n bit code to single active output. The information present in one code is obtained back into the desired code. In general Decoder is circuit which convert specific code into more general form of code. A decoder is a logic circuit that accepts a set of inputs that represents a binary number and activates only the output that corresponds to the input binary number. In other words, a decoder circuit looks at its inputs, determines which binary number is present and activates the one output that corresponds to that number; all other outputs remain inactive.

The basic function of a decoder is used to detect the presence of a specified combination of bits (code) on its inputs and to indicate the presence of that code by a specific output level. In its general form, a decoder has n input lines to handle n bits and from one out of 2^n output lines to indicate the presence of one or more n bit combinations. A decoder is similar to de-multiplexer without Data input. The only inputs are the select inputs or control inputs. As its name indicates, a decoder is a circuit that decodes an input code. Given a binary code of n-bits, a decoder will tell which code is this out of the 2^n possible codes.

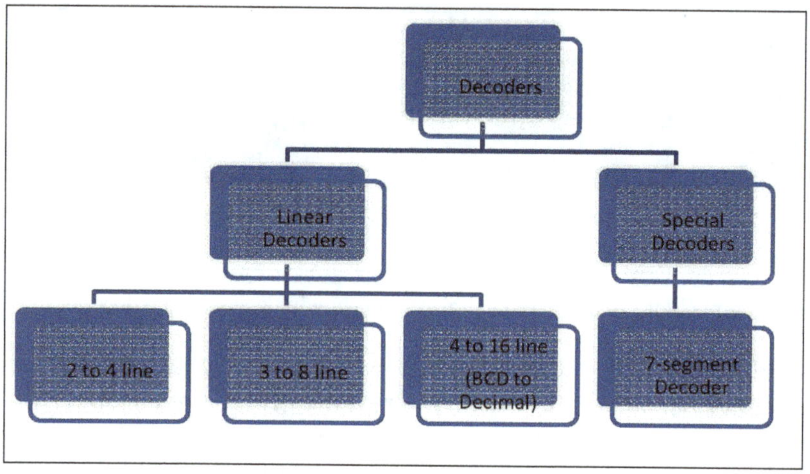

Classification of decoders.

There are two types of decoders: linear or binary decoders and special decoders as shown in figure .In binary (or linear) decoder, only one output line corresponding to input binary code is activated. Whereas the special decoders do not limit activation of single output like binary decoder/demultiplexer but can have more than two outputs activated at a time.

Linear or Binary Decoders

A basic decoder is basically a multiple-input and multiple output logic devices that converts coded inputs into coded output. Here input and output codes are different in type and size. The binary decoder is n-line to 2^n line decoder. These decoders convert binary information from n coded inputs to maximum of 2^n unique outputs. Typical examples are 2 to 4 line decoder, 3 to 8 decoder or 4 to 16 decoder and so on. Thus, the decoder can decode a two, three or four-bit binary number. Decoder may have less than 2^n outputs e.g. BCD to decimal decoder. Table indicates a list of decoder ICs available in TTL logic family.

Table: TTL ICs for Decoder.

IC Number	Description	Output
74138	3 to 8 line decoder/demultiplexer	Inverted output
74139	2 to 4 line decoder/demultiplexerr	Inverted output
74155	2 to 4 line decoder/demultiplexer	Complementary
74156	2 to 4 line decoder/demultiplexer	Open collector
74154	4 to 16 line decoder/demultiplexer	Same as input
74159	4 to 16 line decoder/demultiplexer	Open collector

A 2 to 4 Line Decoder

The simplest decoder is a 2 to 4 line decoder. It has two inputs S_1 and S_0 and four outputs denoted by Y_0, Y_1, Y_2 and Y_3. Figure indicates the logic symbol and Truth table of 2 to 4 line decoder. The output of the decoder is active HIGH. In this decoder, note that S_1 represents MSB and S_0 is the LSB of the input code.

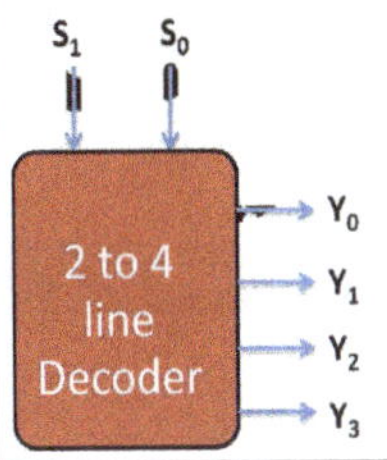

Inputs		Outputs			
S_1	S_0	Y_0	Y_1	Y_2	Y_3
0	0	1	0	0	0
0	1	0	1	0	0
1	0	0	0	1	0
1	1	0	0	0	1

Logic Symbol and Truth Table of 2 to 4 line decoder.

From the truth table of 2 to 4 line decoder, one can obtain the Boolean expression for each output. From these logic expressions, it is possible to draw the logic diagram for 2 to 4 line decoder. The logic diagram of 2 to 4 line decoder is shown in figure below. The logic diagram is generated with the help of AND gates and Inverters. AND gate provides the product terms or MINTERMS of the decoder for input combination.

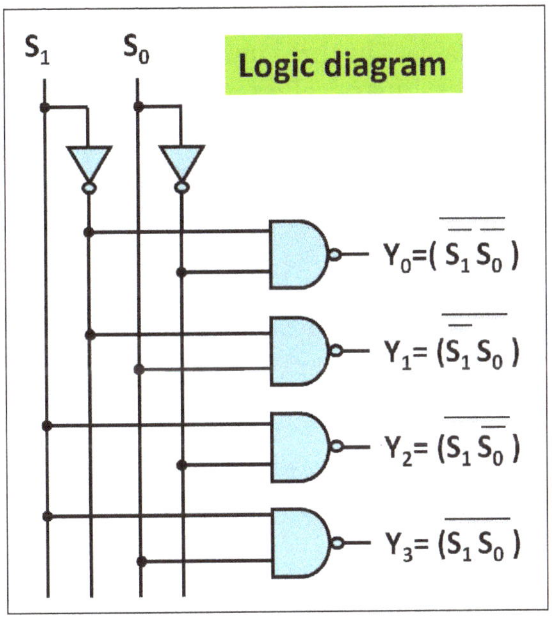

Logic diagram of 2 to 4 line decoder.

Active LOW Decoder

In many applications, decoders are designed with active low output. In such a case AND gates are replaced by NAND gates. Other interconnections with select inputs are unchanged. Figure below shows the active LOW 2 to 4 line decoder.

Generally, decoders have the "enable" input.The enable input performs no logical operation, but is only used for making the decoder ACTIVE or INACTIVE.

- If the enable E is zero, then all outputs are zero regardless of the input values.
- If E is '1', then the decoder performs its normal operation.

Such decoder has active HIGH enable inputs. In many decoders active LOW enable input is also preferred.

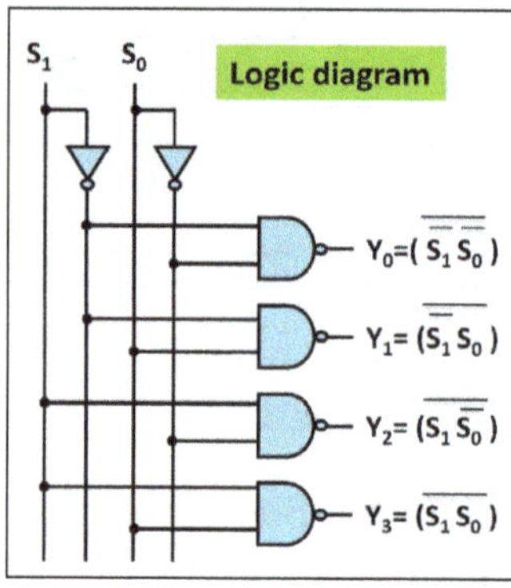

Function Table For active LOW decoder

Inputs		Outputs			
S_1	S_0	Y_0	Y_1	Y_2	Y_3
0	0	0	1	1	1
0	1	1	0	1	1
1	0	1	1	0	1
1	1	1	1	1	0

Active LOW 2 to 4 line decoder.

A 3 to 8 (Binary to Octal) Decoder

In three to eight decoder, there are three inputs and eight outputs. S_0 is the least significant variable, while S_2 is the most significant variable. The three inputs are decoded into eight outputs namely Y_0, Y_1,...Y_7. The function table indicates, binary values at the input from eight combinations activate the corresponding output line. This 3 to 8 line decoder is also known as binary to octal decoder.

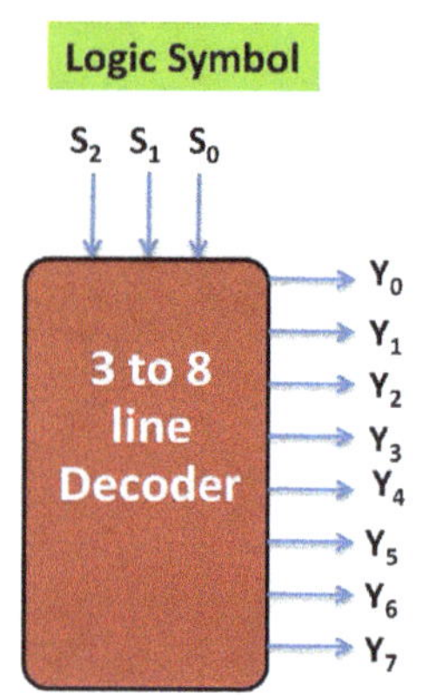

Inputs			Outputs							
S_2	S_1	S_0	Y_0	Y_1	Y_2	Y_3	Y_4	Y_5	Y_6	Y_7
0	0	0	1	0	0	0	0	0	0	0
0	0	1	0	1	0	0	0	0	0	0
0	1	0	0	0	1	0	0	0	0	0
0	1	1	0	0	0	1	0	0	0	0
1	0	0	0	0	0	0	1	0	0	0
1	0	1	0	0	0	0	0	1	0	0
1	1	0	0	0	0	0	0	0	1	0
1	1	1	0	0	0	0	0	0	0	1

Logic symbol and function table of 3 to 8 line decoder.

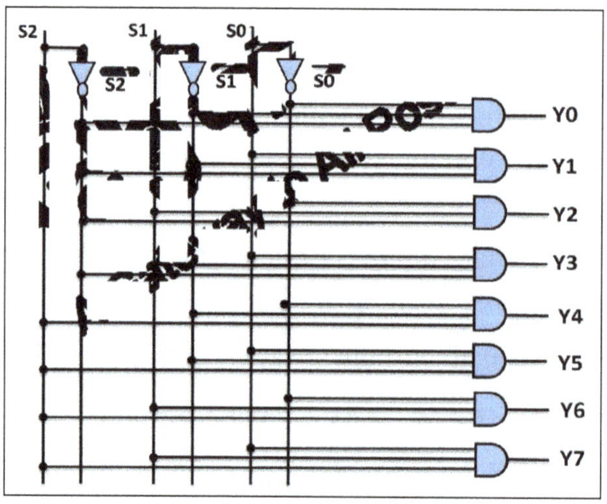

A 3 to 8 (binary to octal) decoder.

In three to eight decoder there are three inputs and eight outputs, as shown in figure A 3 to 8 line decoder generates 8 product terms corresponding to each binary code applied at the three inputs. Thus there is a need of eight 3-input AND gates and 3 NOT gates to generate the complement of input along with un-complemented input. It is possible to verify the truth table by providing input condition and checking whether desired product term is activated on not. In many decoders, there is an enable function, a LOW level on input E1, E2 and a HIGH level on input E3, is required in order to make the enable gate output HIGH and to enable the decoding action.

Input binary code makes only the respective output High and other outputs will be inactive i.e. Low. When select inputs are 000 then the decoder generates only $Y_{0=1}$ and other outputs are 0. From the truth table it is clear that the input binary code decides which output is to be activated.

A 4 to 16 Line (Binary to Hexadecimal) Decoder

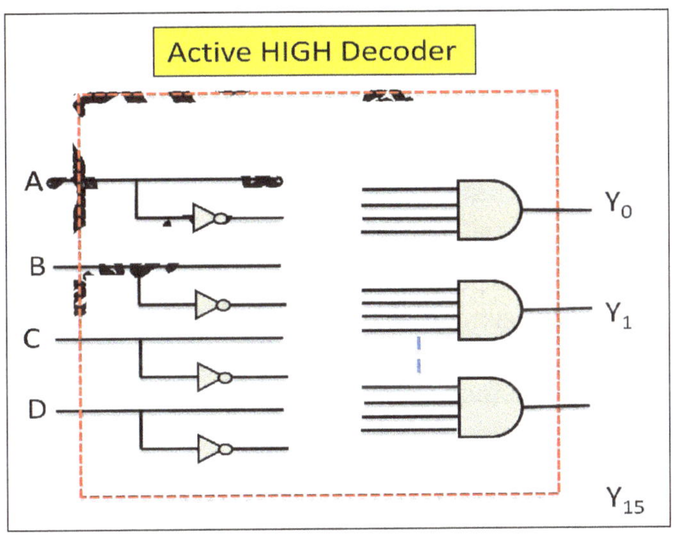

A 4 to 16 decoder.

The 4 to 16 decoder is also popularly known as Binary to Hexadecimal decoder. It converts each binary code into one of sixteen possible outputs as shown in figure. The method of implementation is essentially the same as previous decoders. This decoder has 4 inputs say A, B, C and D. The 4 bit binary activates only one out of 16 possible output product terms, which are labeled as Y_0, Y_1, ... Y_{15}. All the outputs are active HIGH in this implementation. The design implementation requires sixteen 4-input AND gates and 4 NOT gates.

Let us now look at the scaled down version of the previous decoder. The BCD to decimal decoder converts each binary code into one of ten possible outputs indicating decimal number. The method of implementation is essentially the same as previous decoders. This decoder has 4 inputs say A, B, C and D. This 4 bit binary activates only one out of 16 possible product terms, which are labeled as Y_0, Y_1, ... Y_9. Please note that All the outputs are active LOW in this implementation. This implementation requires Ten 4-input NAND gates and 4 NOT gates. Active LOW outputs in the decoder are implemented using NAND gates and active HIGH outputs are implemented using AND gates. All decoder/driver ICs have active HIGH inputs and active LOW outputs. The outputs could be either totem pole or pen collector form. These ICs can be used either as simple indicator or driver for nixie tube or LED or Relay.

Table: TTL ICs for BCD to Decimal Decoder.

IC No	Description	Output	Applications
7441	BCD To Decimal Decoder	Open Collector Output	Nixie Driver
7442	BCD To Decimal Decoder	Totem Pole Output	LED Driver
7445	BCD To Decimal Decoder	Open Collector Output	Indicator/ Relay Driver
74141	BCD To Decimal Decoder	Open Collector Output	Nixie Tube Driver
74145	BCD To Decimal Decoder	Open Collector Output	Indicator/Relay Driver
74445	BCD To Decimal Decoder	Open	Indicator/Relay Driver

Special Decoders

The binary decoders studied so far generates only one output of many in response to the binary inputs to the decoder. Whereas special decoder activates more than one outputs simultaneously for the applied input binary code. Many digital types of equipment require some means for displaying information. One of the simplest and most popular methods for displaying numeric digits utilizes seven segment configuration to for numbers 0 to 9 and sometimes the hex characters A to F.

Special decoders are designed to drive the Seven Segment display. Each segment is made up of material that emits light when current is passed through it. Most commonly used devices include light emitting diode (LEDs). The seven segment display has 7 LEDs for the segments and one additional LED for dot i.e. used as decimal point. There are two configurations for seven segment display. These are Common Anode and Common Cathode Display.

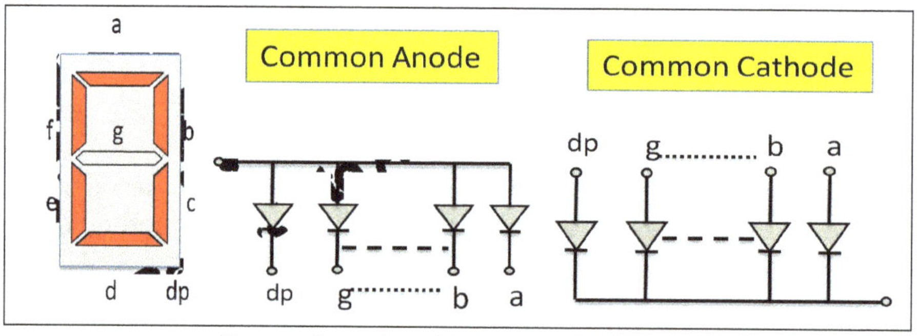

Two configuration of Seven segment display.

In common anode displays, the anodes of all LEDs are connected in common and cathodes are labeled as a, b, c, d, e, f, g, and dp in sequential order. The common anode requires the driving circuit to provide LOW level voltage in order to activate a given segment. When a LOW level is applied to a segment input, the LED is forward biased and current flows through it.

BCD to Seven Segment Decoder

Designer can select either common anode or common cathode depending on the application. The logic symbol of BCD to seven segment decoder/driver has 4 inputs and eight outputs are shown in figure. This BCD to seven segment decoder generates a seven segment code in response to binary code for driving the seven segment display.

In order to display a decimal digit, it necessary for the decoder to activate the desired segment. Let us now prepare a table for displaying a decimal digit and corresponding activated segments. For displaying digit 0, LED segments a, b, c, d, e and f must be activated and segments g and dp are OFF.

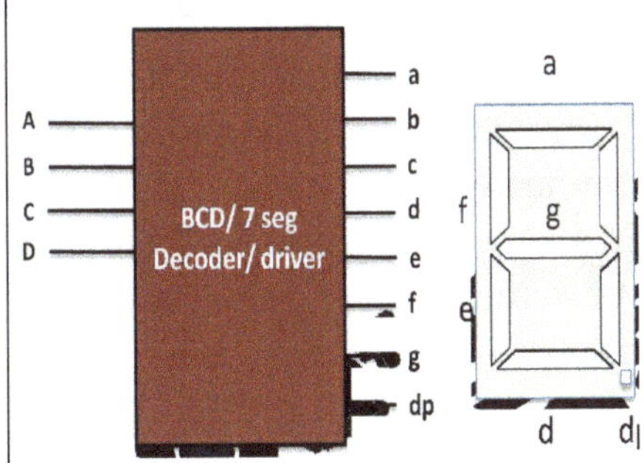

Digit	Segments Activated
0	a,b,c,d,e,f
1	b,c
2	a,b,g,e,d
3	a,b,g,c,d
4	b,c,f,g
5	a,c,d,f,g
6	a,c,d,e,f,g
7	a,b,c
8	a,b,c,d,e,f,g
9	a,b,c,f,g

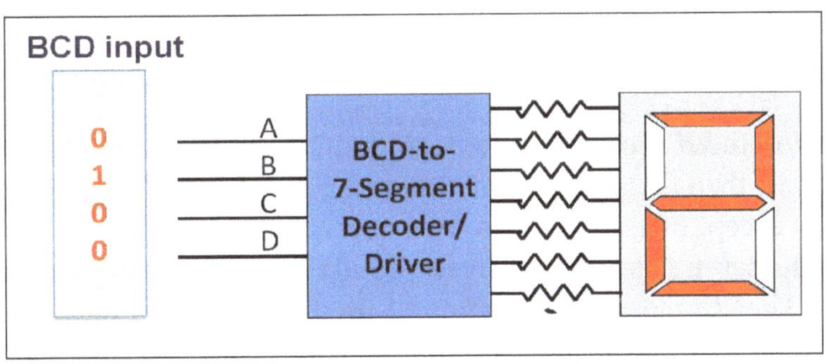

Interfacing seven segment display.

Let us now find out activated segments for displaying digit 1. Here only b and c segments must be activated and other segments are deactivated. For displaying digit 2, LED segments a, b, g, e, d are activated and so on. Let us now study the actual interfacing of BCD to seven segment decoder/driver with the seven segment display. Here limiting resistors are connected between them. The resistor value decides the brightness of the segment. Let us now consider display of decimal number 2. If BCD input is 0010 then decoder generates outputs to display decimal number 2. And so on.

Table: BCD to seven segment decoder/driver ICs.

IC No	Output	Rating (Maximum Voltage, Sink Current)
7446	Active low, open collector	30V, 40 mA
7447	Active low, open collector	15V, 40 mA
7448	Active High, Pull Up Resistor	5.5V, 6 mA
7449	Active high, open collector	5.5V, 8 mA

Let us look at the commercially available BCD to seven segment decoder/driver ICs. IC 7446 and 7447 has open collector output and the decoder requires active LOW output to drive the common anode display configuration of 7 segment display. Whereas IC 7448 is recommended for driving common cathode displays.

In addition to 4 BCD inputs and seven segment outputs, this decoder has 3 additional pins. These are used for special purpose. The function of LT i.e. Lamp test terminal is to test all the seven segments of the display. As it is active LOW input, LOW on this pin turns ON all the LED segments of the display. In normal decoding application, this terminal should be connected to logic HIGH. Let us now talk about BI i.e. Blanking Input. If this terminal is connected to LOW level, the display is switched OFF irrespective of BCD inputs. This is used for conserving the power in multiplexed display application.

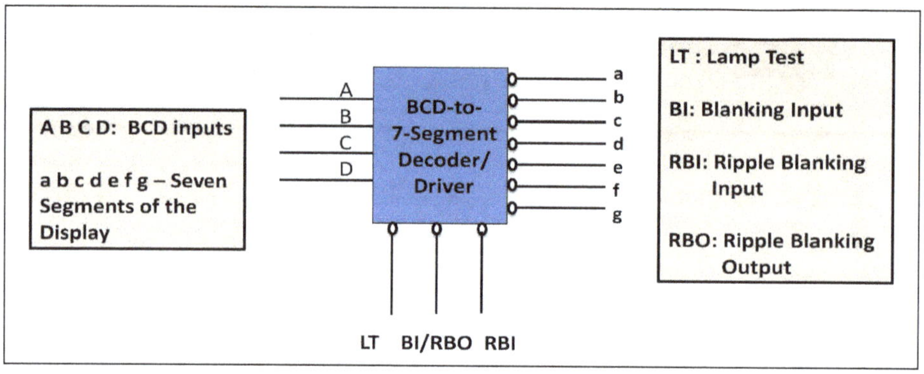

: BCD to seven segment decoder/driver ICs.

RBI i.e. Ripple Blanking Input is connected to logic "HIGH" in normal decoding application. If it is connected to LOW logic, then the segment outputs will generate the data for normal seven segment decoding applications except BCD input Zero. Whenever all the BCD inputs are zero corresponding to decimal zero then the seven segment display is turned OFF. This is used for Zero blanking in multi digit display. The RBO i.e. Ripple Blanking Output is used for cascading purpose and is connected to RBI terminal of succeeding stage. Normally there is a common terminal for BI and RBO in many ICs.

IC 7446/47 is a TTL IC for driving a common anode display as shown in this figure. Common anode terminal is connected to +Vcc. Decoder IC has open collector outputs where seven segments are connected through the current limiting resistors. Logic circuit inside the IC converts BCD input to a required output.

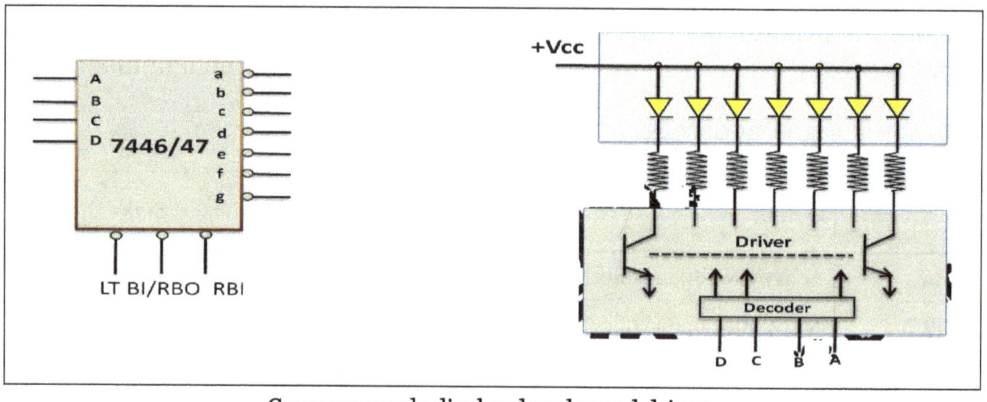

Common anode display decoder and driver.

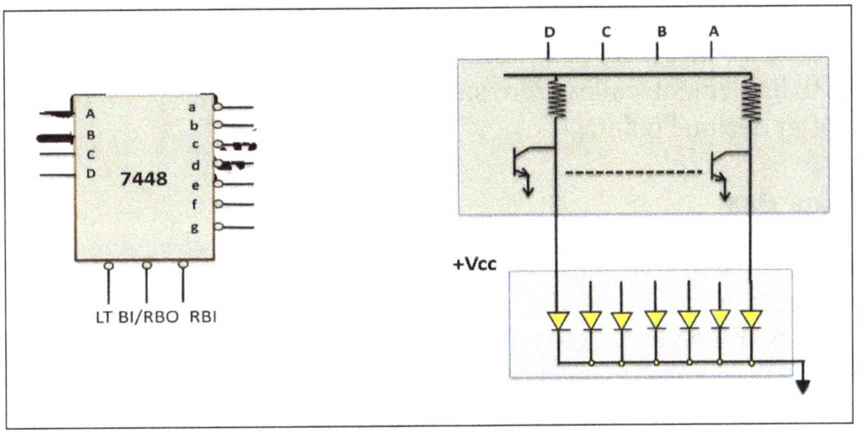

Common cathode display decoder and driver.

IC 7448 is a TTL IC for driving a common cathode display as shown in this figure. Common cathodes of all LED segments are connected to common ground. The decoder IC has its own current limiting resistor on the chip. Logic circuit inside the IC converts BCD input to a required output.

Applications of Decoder

There are many applications of Decoder like leading zero suppression, multiplexed display, time division multiplexing, data decoding and function generator.

Leading Zero Suppression

First application of decoder is in leading zero suppression. Many seven segment decoders have built in capacity of zero suppression. This facility is useful in multi digit display. This technique is used to blank out unnecessary zeros in the display. For example, number 35 is to be displayed on 4 digit seven segment display. In place of displaying actually 0035 leading zeros will be blanked out in the displayed. It may help user to read the number easily.

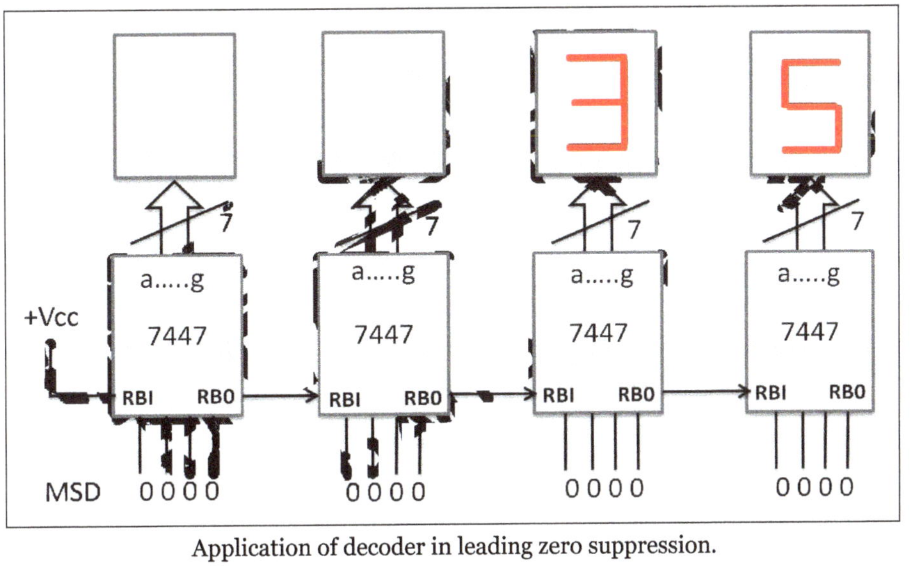

Application of decoder in leading zero suppression.

In this circuit, RBI of most significant display is connected to =Vcc and the RBO is connected to RBI of the next decoder. This is the way how ripple blanking is implemented through cascading. It is also possible to implement trailing zero suppression in which zeros after the number are suppressed, usually after decimal point.

Multiplexed Display

Next application of the decoder is in Multiplexed display logic. In this example, single seven segment display decoder can be used to display 2 digit data. Two BCD numbers are applied to multiplexer inputs. When data select line is LOW, first BCD number is passed to the seven segment decoder and generates 7-seven segment code at the output of the decoder. In addition the IC 74139, the linear decoder enables the first display to accept data from seven segment decoder.

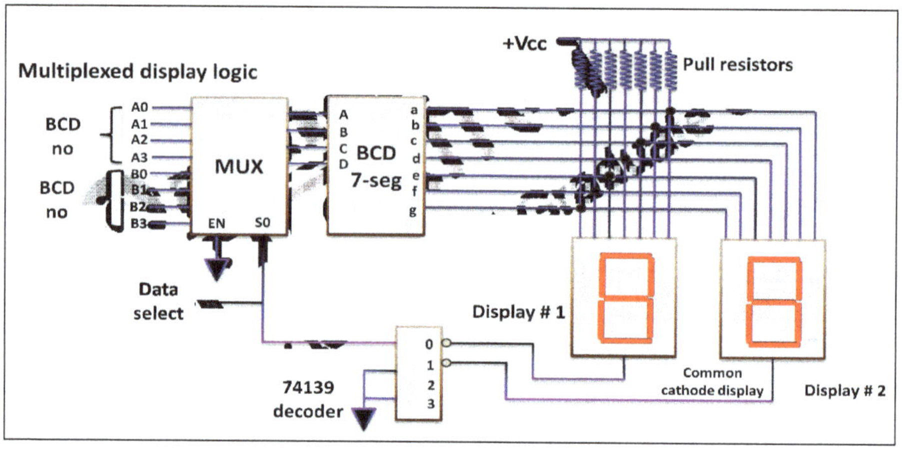

Application of decoder in Multiplexed display logic.

When data select line is HIGH, Second BCD number is passed to the seven segment decoder and generates 7-seven segment code at the output of the decoder. In addition, the IC 74139, the linear decoder enables the second display to accept data from seven segment decoder. In this example, both binary and seven segment decoders are used.

Time Division Multiplexing

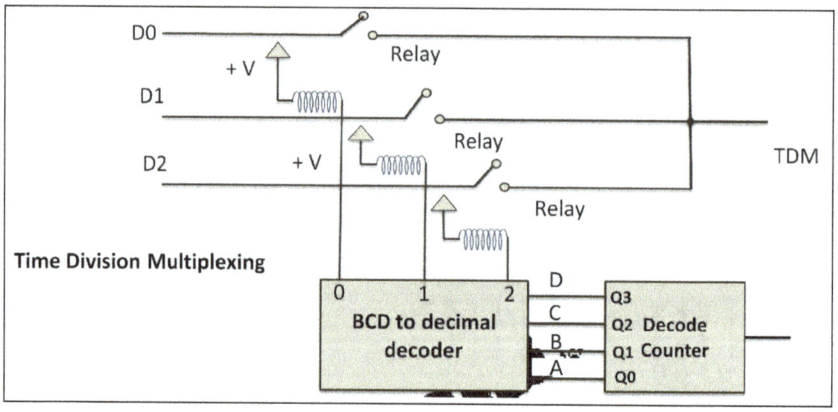

Application of decoder in Time division multiplexing.

Let us understand the use of decoder in time division multiplexing and also for timing and sequencing. The TDM accepts data from many sources and sends a selected data to the output for transmission. In this case output is generated in a time multiplexed manner i.e. one at a time depending on the decoder input coming from a counter. The counter is continuously pulsed and decoder outputs are activated sequentially A, B, C,....etc. This activates one relay at a time to connect the data source to the output.

Address Decoding for I/O Devices

A simplified computer I/O port system with a port address decoder with only four address lines. Decoders are used in many types of applications. One example is in computers for I/O selection. Computer must communicate with a variety of external devices called peripherals by sending and/or receiving data through what is known as input/output (I/O) ports. Each I/O port has a number, called an address, which uniquely identifies it. When the computer wants to communicate with a particular device, it issues the appropriate address code for the I/O port to which that particular device is connected. The binary port address is decoded and appropriate decoder output is activated to enable the I/O port. Binary data are transferred within the computer on a data bus, which is a set of parallel lines.

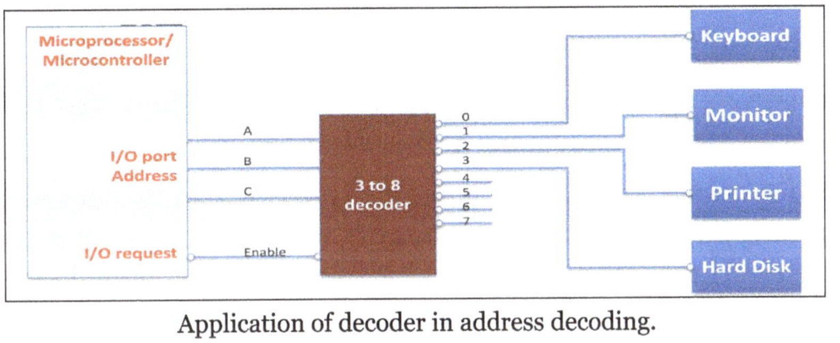

Application of decoder in address decoding.

Function Generator

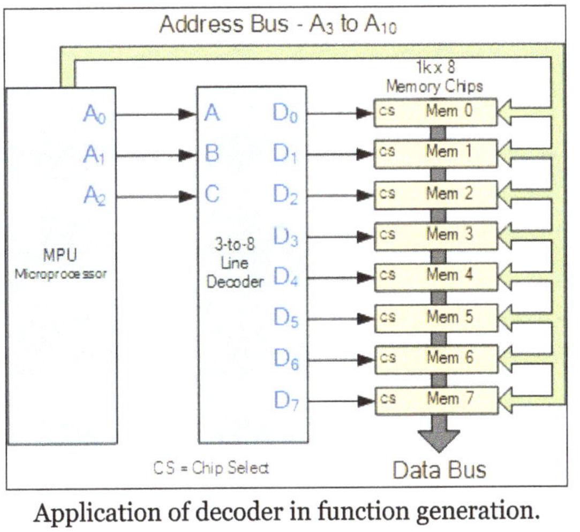

Application of decoder in function generation.

Any combinational logic circuit with n inputs and m outputs can be implemented with an n-to-2^n-line decoder and m OR gates. Procedure:

- Express the given Boolean function in sum of min-terms.

- Choose a decoder to generate all the min-terms of the input variables.
- Select the inputs to each OR gate from the decoder outputs according to the list of min-term for each function.

For adder circuit:

$$S(x,y,z) = \Sigma m(1,2,4,7)$$
$$C(x,y,z) = \Sigma m(3,5,6,7)$$

where $\sum$ indicates sum, m indicates min-term and the number in brackets indicate the decimal equivalent. Since there are three inputs and a total of eight min-terms, we need a 3-to-8 line decoder. The decoder generates the eight min-terms for x, y, z. The OR gate for output S forms the logical sum of min-terms 1, 2, 4, and 7. The OR gates for output C forms the logical sum of min-terms 3, 5, 6, and 7. There are many other applications of decoders like: Line coding and decoding for digital data communication, Speech coding and decoding, Constructing higher order decoder and memory addressing.

Multiplexers and Demultiplexers

In large scale digital systems, a single line is required to carry on two or more digital signals – and of course at a time, one signal can be placed on one line. But, what is required is a device that will allow us to select; and, the signal we wish to place on a common line, such a circuit is referred to as a multiplexer. The function of a multiplexer is to select the input of any 'n' input lines and feed that to one output line. The function of a demultiplexer is to inverse the function of the multiplexer. The shortcut forms of the multiplexer and demultiplexers are mux and demux. Some multiplexers perform both multiplexing and demultiplexing operations. The main function of the multiplexer is that it combines input signals, allows data compression, and shares a single transmission channel.

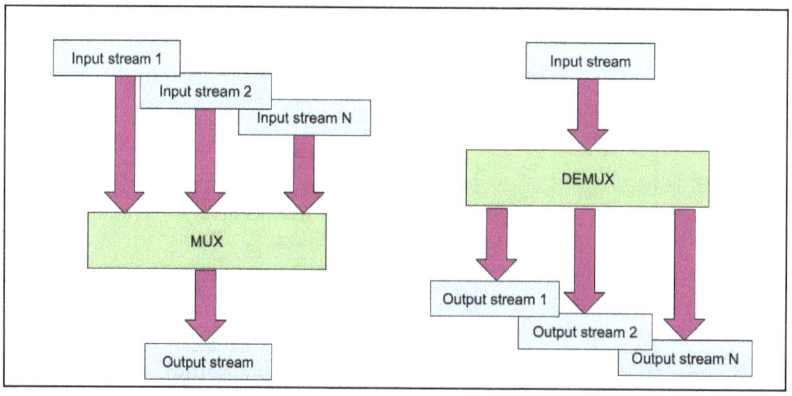

In-network transmission, both the multiplexer and demultiplexer are combinational circuits. A multiplexer selects an input from several inputs then it is transmitted in the form of a single line. An alternative name of the multiplexer is MUX or data selector. A demultiplexer uses one input signal and generates many. So it is known as Demux or data distributor.

Multiplexer

The multiplexer is a device that has multiple inputs and single line output. The select lines determine which input is connected to the output, and also increase the amount of data that can be sent over a network within a certain time. It is also called a data selector.

The single-pole multi-position switch is a simple example of a non-electronic circuit of the multiplexer, and it is widely used in many electronic circuits. The multiplexer is used to perform high-speed switching and is constructed by electronic components.

Multiplexers are capable of handling both analog and digital applications. In analog applications, multiplexers are made up of relays and transistor switches, whereas in digital applications, the multiplexers are built from standard logic gates. When the multiplexer is used for digital applications, it is called a digital multiplexer.

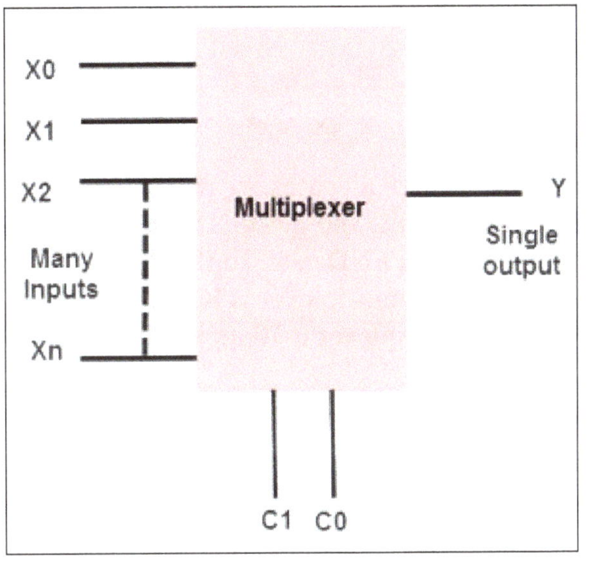

Multiplexer Types

Multiplexers are classified into four types:

- 2-1 multiplexer (1 select line).
- 4-1 multiplexer (2 select lines).
- 8-1 multiplexer (3 select lines).
- 16-1 multiplexer (4 select lines).

4-to-1 Multiplexer

The 4X1 multiplexer comprises 4-input bits, 1- output bit, and 2- control bits. The four input bits are namely 0, D1, D2, and D3, respectively; only one of the input bits is transmitted to the output. The o/p 'q' depends on the value of control input AB. The control bit AB decides which of the i/p data bit should transmit the output. The following figure shows the 4X1 multiplexer circuit diagram using AND gates. For example, when the control bits AB = 00, then the higher AND gates are allowed while remaining AND gates are restricted. Thus, data input D0 is transmitted to the output 'q".

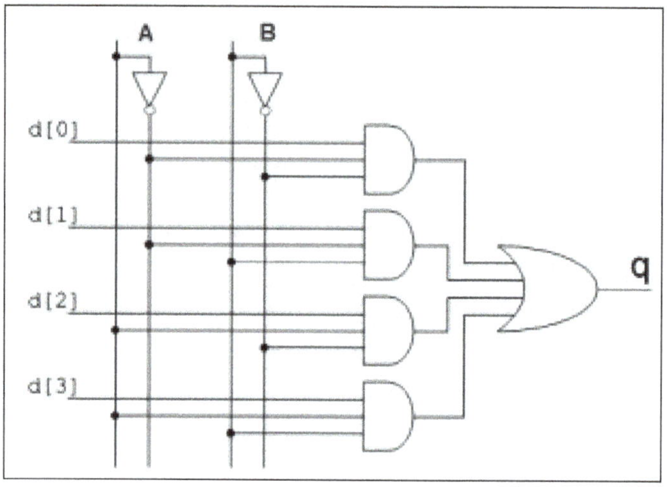

4X1 Mux.

If the control input is changed to 11, then all gates are restricted except the bottom AND gate. In this case, D3 is transmitted to the output, and q=D0. If the control input is changed to AB = 11, all gates are disabled except the bottom AND gate. In this case, D3 is transmitted to the output, and q = D3. The best example of a 4X1 multiplexer is IC 74153. In this IC, the o/p is the same as the i/p. Another example of a 4X1 multiplexer is IC 45352. In this IC, the o/p is the compliment of the i/p.

8-to-1 Multiplexer

The 8-to-1 multiplexer consists of 8 input lines, one output line, and 3 selection lines.

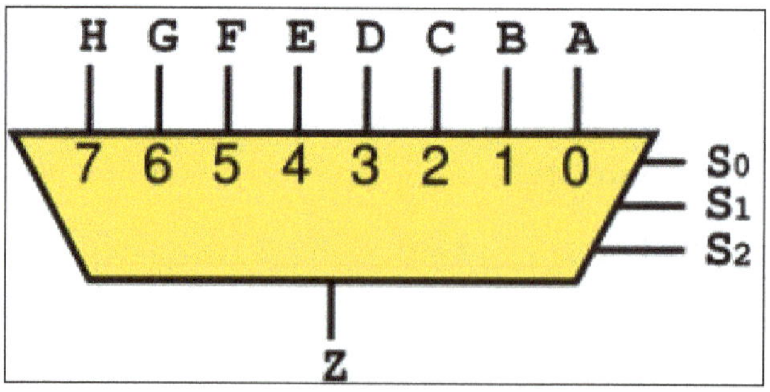

8-to-1 Mux.

8-1 Multiplexer Circuit

For the combination of a selection input, the data line is connected to the output line. The circuit shown below is an 8*1 multiplexer. The 8-to-1 multiplexer requires 8 AND gates, one OR gate, and 3 selection lines. As an input, the combination of selection inputs is giving to the AND gate with the corresponding input data lines.

In a similar fashion, all the AND gates are given connection. In this 8*1 multiplexer, for any selection line input, one AND gate gives a value of 1 and the remaining all AND gates give 0. And, finally, by using OR gates, all the AND gates are added; and, this will be equal to the selected value.

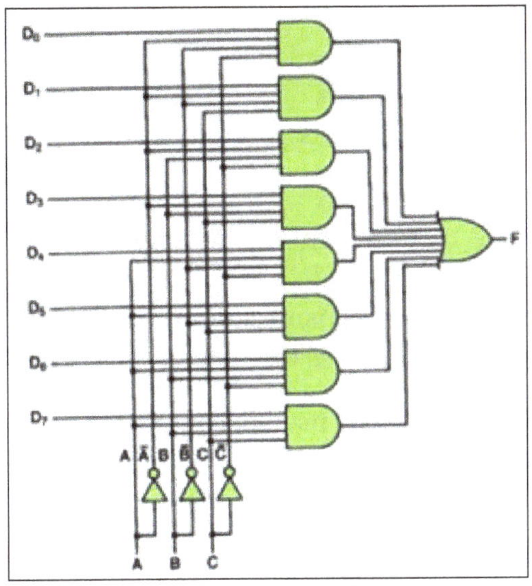

8-to-1 Mux Circuit.

Advantages and Disadvantages of Multiplexer

The advantages of multiplexer include the following:

- In multiplexer, the usage of a number of wires can be decreased.
- It reduces the cost as well as the complexity of the circuit.
- The implementation of a number of combination circuits can be possible by using a multiplexer.
- Mux doesn't require K-maps & simplification.
- The multiplexer can make the transmission circuit less complex & economical.
- The dissipation of heat is less because of the analog switching current which ranges from 10mA to 20mA.
- The multiplexer ability can be extended to switch audio signals, video signals, etc.
- The digital system reliability can be improved using a MUX as it decreases the number of exterior wired connections.
- MUX is used to implement several combinational circuits.
- The logic design can be simplified through MUX.

The disadvantages of multiplexer include the following:

- Additional delays required within switching ports & I/O signals which propagate throughout the multiplexer.
- The ports which can be utilized at the same time have limitations.

- Switching ports can be handled by adding the complexity of firmware.
- The controlling of multiplexer can be done by using additional I/O ports.

Applications of Multiplexers

Multiplexers are used in various applications wherein multiple-data need to be transmitted by using a single line.

- Communication System: A communication system has both a communication network and a transmission system. By using a multiplexer, the efficiency of the communication system can be increased by allowing the transmission of data, such as audio and video data from different channels through single lines or cables.
- Computer Memory: Multiplexers are used in computer memory to maintain a huge amount of memory in the computers, and also to reduce the number of copper lines required to connect the memory to other parts of the computer.
- Telephone Network: In telephone networks, multiple audio signals are integrated on a single line of transmission with the help of a multiplexer.
- Transmission from the Computer System of a Satellite: The multiplexer is used to transmit the data signals from the computer system of a spacecraft or a satellite to the ground system by using a GSM satellite.

Demultiplexer

Demultiplexer is also a device with one input and multiple output lines. It is used to send a signal to one of the many devices. The main difference between a multiplexer and a de-multiplexer is that a multiplexer takes two or more signals and encodes them on a wire, whereas a demultiplexer does reverse to what the multiplexer does.

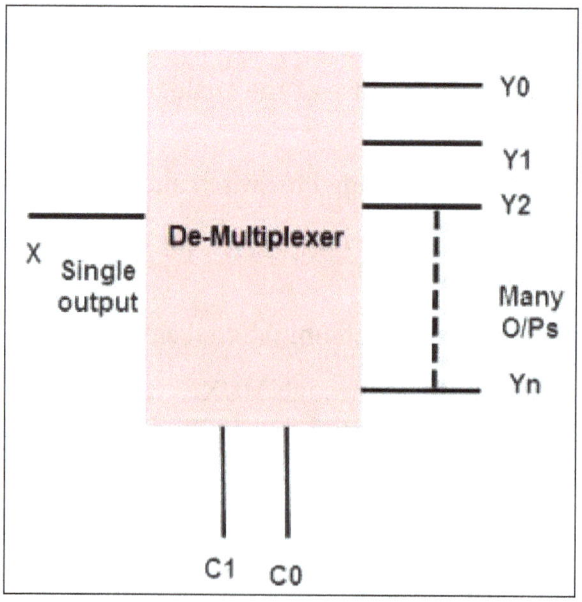

Types of Demultiplexer

Demultiplexers are classified into four types:

- 1-2 demultiplexer (1 select line).
- 1-4 demultiplexer (2 select lines).
- 1-8 demultiplexer (3 select lines).
- 1-16 demultiplexer (4 select lines).

1-4 Demultiplexer

The 1-to-4 demultiplexer comprises 1-input bit, 4-output bits, and control bits. The 1X4 demultiplexer circuit diagram is shown below:

The i/p bit is considered as Data D. This data bit is transmitted to the data bit of the o/p lines, which depends on the AB value and the control i/p. When the control i/p AB = 01, the upper second AND gate is permitted while the remaining AND gates are restricted. Thus, only data bit D is transmitted to the output, and Y_1 = Data.

If the data bit D is low, the output Y_1 is low. IF data bit D is high, the output Y_1 is high. The value of the output Y_1 depends upon the value of data bit D, the remaining outputs are in a low state. If the control input changes to AB = 10, then all the gates are restricted except the third AND gate from the top. Then, data bit D is transmitted only to the output Y_2; and, Y_2 = Data. The best example of 1X4 demultiplexer is IC 74155.

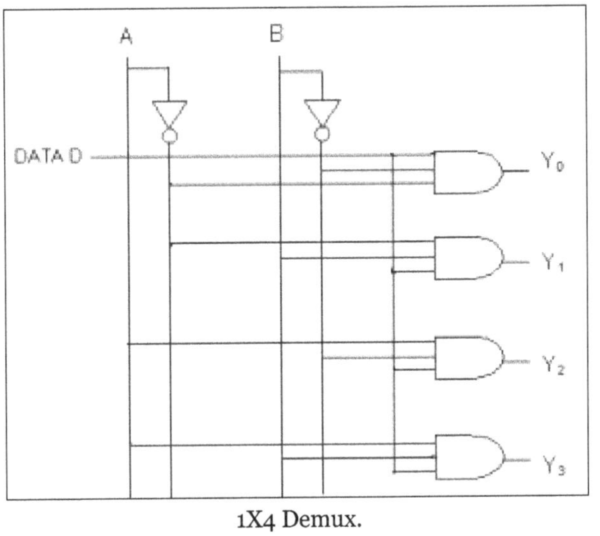

1X4 Demux.

1-8 Demultiplexer

The demultiplexer is also called a data distributor as it requires one input, 3 selected lines, and 8 outputs. De-multiplexer takes one single input data line and then switches it to any one of the output lines. The 1-to-8 demultiplexer circuit diagram is shown below; it uses 8 AND gates for achieving the operation.

The input bit is considered as data D and it is transmitted to the output lines. This depends on the control input value of the AB. When AB = 01, the upper second gate F_1 is enabled, while the remaining AND gates are disabled, and the data bit is transmitted to the output giving F_1= data. If D is low, the F_1 is low, and if D is high, the F_1 is high. So the value of the F_1 depends on the value of D, and the remaining outputs are in the low state.

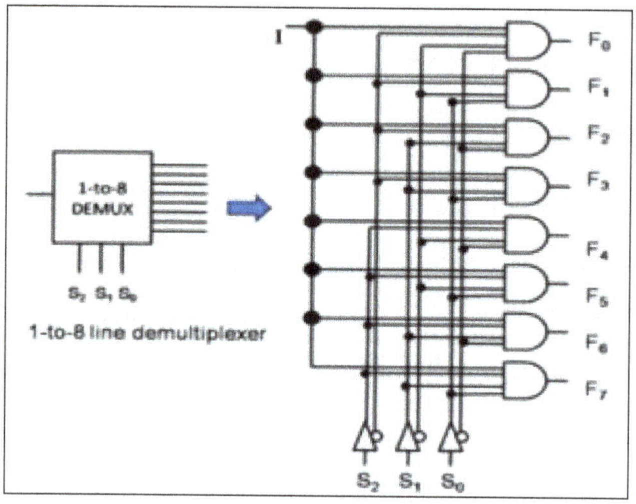

1-8 Demux Circuit.

Advantages and Disadvantages of Demultiplexer

The advantages of demultiplexer include the following:

- A demultiplexer or Demux is used to divide the mutual signals back into separate streams.
- The function of Demux is quite opposite to MUX.
- The Audio or Video signals transmission needs a combination of Mux and Demux.
- Demux is used as a decoder within the security systems of banking sectors.
- The communication system efficiency can be enhanced through the combination of Mux & Demux.

The disadvantages of demultiplexer include the following:

- Bandwidth wastage might happen.
- Because of the synchronization of the signals, delays might take place.

Applications of Demultiplexer

Demultiplexers are used to connect a single source to multiple destinations. These applications include the following:

- Communication System: Mux and demux both are used in communication systems to carry out the process of data transmission. A De-multiplexer receives the output signals from the multiplexer and at the receiver end it converts them back to the original form.

- Arithmetic Logic Unit: The output of the ALU is fed as an input to the De-multiplexer, and the output of the demultiplexer is connected to multiple registers. The output of the ALU can be stored in multiple registers.
- Serial to Parallel Converter: This converter is used to reconstruct parallel data. In this technique, serial data is given as an input to the De-multiplexer at a regular interval, and a counter is attached to the demultiplexer at the control input to detect the data signal at the output of the demultiplexer. When all data signals are stored, the output of the demux can be read out in parallel.

Difference between Multiplexer and Demultiplexer

Multiplexer	Demultiplexer
A multiplexer (Mux) is a combinational circuit that uses several data inputs to generate a single output.	A demultiplexer (Demux) is also a combinational circuit that uses single input that can be directed throughout several outputs.
Multiplexer includes several inputs and the single output.	Demultiplexer includes single input and several outputs.
A multiplexer is a data selector.	The demultiplexer is a data distributor.
It is a digital switch.	It is a digital circuit.
It works on the principle of many to one.	It works on the principle of one-to-many.
The parallel to serial conversion is used in the multiplexer.	The serial to parallel conversion is used in Demultiplexer.
The multiplexer used in TDM (Time Division Multiplexing is at the end of the transmitter.	The demultiplexer used in TDM (Time Division Multiplexing is at the end of the receiver.
The multiplexer is called MUX.	The demultiplexer is called Demux.
It doesn't use any extra gates while designing.	In this, additional gates are necessary while designing demux.
In Multiplexer, control signals are used to choose the specific input that has to be sent at the output.	Demultiplexer uses the control signal to permit us to include several outputs.
The multiplexer is used to improve the efficiency of the communication system using transmission data like transmission of audio as well as video.	Demultiplexer gets the o/p signals from the Mux & changed them to the unique form at the end of the receiver.
The different types of multiplexers are 8-1 MUX, 16-1 MUX, and 32-1 MUX.	The different types of demultiplexers are 1-8 Demux, 1-16 Demux, 1-32 Demux.
In multiplexer, the set of selection lines are used to control the specific input.	In demultiplexer, the selection of output line can be controlled through n-selection lines bit values.

Key Difference between Multiplexer and Demultiplexer

- The combinational logic circuits like multiplexer and demultiplexer are used within communication systems however their function is accurately opposite to each other because one works on multiple inputs whereas the other works on only input.
- Multiplexer or Mux is an N-to-1 device whereas demultiplexer is a 1-to-N device.

- A multiplexer is used to convert several analog or digital signals into a single o/p signal through different control lines. These control lines can be determined by using this formula like 2n=r where 'r' is the no of i/p signals & 'n' is the no of required control lines.
- The data conversion method used in MUX is parallel to serial and it is not difficult to understand because it uses different inputs. However, DEMUX works quite reverse to MUX like a serial to parallel conversion. So, the number of outputs can be achieved in this case.
- A demultiplexer is used to convert one i/p signal to several. The number of control signals can be determined by using the same formula of MUX.
- Both the Mux and Demux are used to transmit the data over a network in less bandwidth. But multiplexer is used at the transmitter end whereas the Demux is used at the receiver end.

Arithmetic Logic Unit

An arithmetic logic unit (ALU) is a digital system that performs arithmetic and logical operations. ALU is a fundamental building block of a central processing unit (CPU) or a microprocessor of a computer that performs arithmetic (Addition, Subtraction....) and logical operations (AND, OR, NOT ...). The concept of ALU was proposed by mathematician John Von Neumann. ALU is basically a combinational circuit that performs arithmetic and logical operations on binary integers. Whereas many applications demand processing of floating point numbers. Such unit is called floating point unit (FPU), which operates on floating point numbers. Some of the ALUs are also used for processing graphics. Such ALUs require a special unit called as floating point unit (FPU). Sometimes the FPU is used as partner chip called as "coprocessor".

When a floating point operations needs to be performed, the CPU sends it off to the coprocessor. Coprocessor after performing the operation sends back the result to CPU. Presently, there are modern CPUs with an internal FPU. The operations of a microprocessor are performed by one or more ALUs. In a processor, ALU is divided into an AU and LU. The AU performs the arithmetic operations and the LU performs the logical operations.

The arithmetic logic unit (ALU) is designed to be a part of computer CPU or a microprocessor for performing arithmetic and logic operations. In many cases, an ALU is supposed to handle either finite field arithmetic or with allowed range infinite field arithmetic. This indicates use of integer ALU. Present superscalar processors have a separate FPU (floating point unit) and integer ALU. The CPU of the computer loads data from input registers to an ALU. The control unit of a processor tells ALU what of operation to perform on that data. After the operation control unit transfers result of ALU into output register.

Figure above indicates a typical logic symbol of an ALU. In this figure, operands A and B are the two inputs of an ALU. F is another type of input indicating the arithmetic or logic function to be performed on the operands. Y indicates the result of the operation and S indicates status of the ALU result (overflow, negative, carry out....) after the operation.

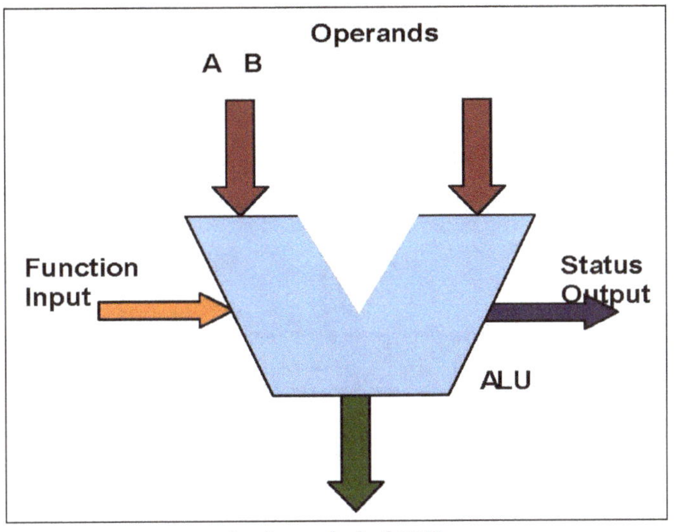

Logic symbol of ALU.

An ALU performs basic arithmetic and logical operations. The arithmetic operations usually implemented are addition, subtraction, multiplication, and division. Similarly, the logic operations are NOT, AND, OR etc. are normally implemented. Along with the result, a set of status flags are set which include Zero, Carry, and Negative or overflow to indicate the status of the result. The arrow symbols indicate more number of bits is available at inputs as well as output. The ALU can handle integers of some data width, which decides the processing ability of the processor. For example 8 bit ALU handles 8-bit data at the inputs. An ALU can be programmed to perform any series of complex arithmetic or logical calculations. The computational capacity of the ALU decides the power of the CPU of a computer.

1-bit ALU Design

An ALU is the fundamental unit of any computing system. Understanding how an ALU is designed and how it works is essential to building any advanced logic circuits. Using this knowledge and experience, we can move on to designing more complex integrated circuits. The goal is to design an n-bit ALU.

Designing n-bit ALU is similar to the way we design n-bit adder. The design begins with building a 1-bit ALU version. This one bit ALU can be built from logic gates, adder and multiplexer. Such 1-bit ALU can be combined together to get n-bit ALU. Let us first build the logic unit first and then combine arithmetic block in it to make an ALU. Let us assume that the unit can handle two 1 bit inputs to produce a required output based on the function selector line. Figure above indicates a simple block for 1-bit logic unit.

Actually there are relatively few basic arithmetic and logic operations needed. Many operations can be implemented in terms of the basic ones. The basic building blocks of a simple logic unit are NOT gate, AND gate, OR gate, XOR gate and a multiplexer as shown in figure.

The simple logic unit contain four basic logic gates to provide four basic logic operations like NOT (complement), AND, OR and XOR. To select one of four logic operations a selector is required. In this unit a simple 4 to 1 multiplexer is preferred for designing of logic unit. To select one of 4 inputs, there are two select inputs S1 and So as shown in table. The function table assumes that both the data inputs A, B are present. The values of select lines decide the nature of operation to be performed on the operands.

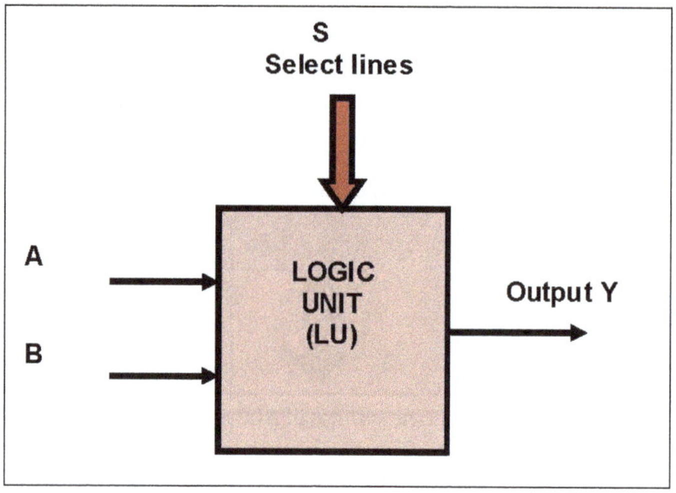

A simple logic unit.

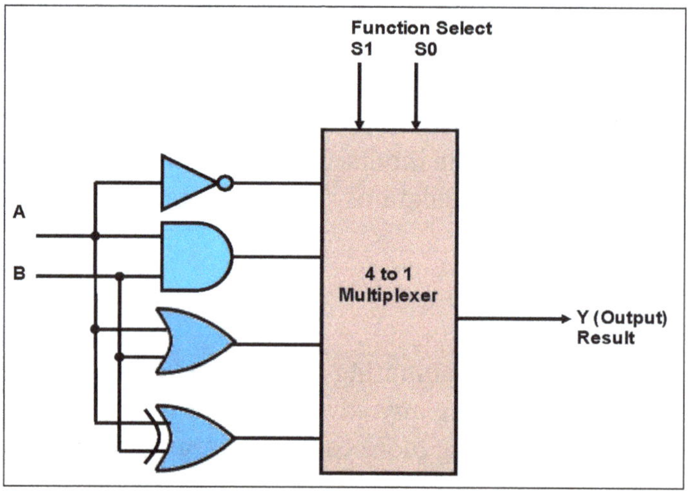

Logic diagram for 1-bit Logic Unit.

Table: Function table of Logic Unit.

S1	So	Output	Operation
0	0	Y=A'	Complement
0	1	Y=A. B	AND
1	0	Y=A + B	OR
1	1	Y= A ⊕ B	XOR

After designing the 1-bit logic unit, the next obvious step is to incorporate an adder block to logic unit. Addition of a full adder allows the unit to perform addition as well. Figure below shows indicates1-bit logic unit with addition feature.

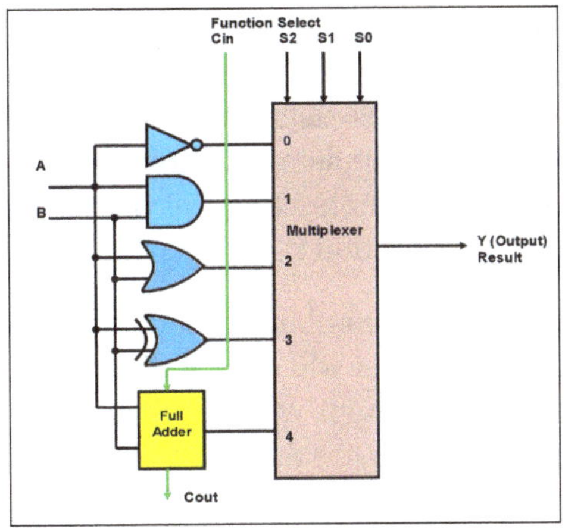

1-bit logic unit with addition.

To select one of four logic operations and addition, multiplexer with 5 inputs is required. In this unit a simple 8 to 1 multiplexer is preferred for designing of logic unit with adder. To select one of 5 inputs, there are three select inputs S2, S1 and So. The function table assumes that both the data inputs A, B are present. The full adders have three data inputs A, B and Ci with sum and carry as outputs. The values of select lines decide the nature of operation to be performed on the operands. Function table as shown table below indicates the correlation between the values of select inputs and the nature of operation i.e. arithmetic or logical operation.

Table: Function table of logic unit with addition.

S2	S1	So	Output	Operation
0	0	0	Y=A'	Complement
0	0	1	Y=A. B	AND
0	1	0	Y=A + B	OR
0	1	1	Y= A ⊕ B	XOR
1	X	X	Y= Sum(A,B) & Co	Addition

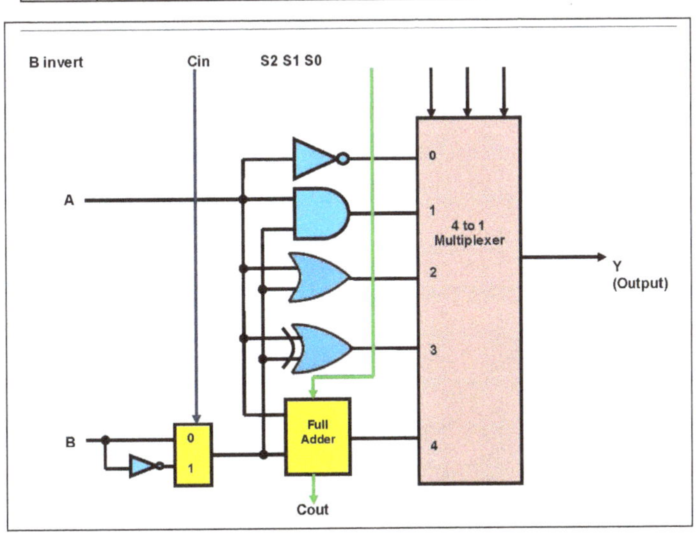

1-bit Arithmetic and logic unit.

For performing subtraction, it is necessary to perform A-B. This subtraction can also be represented as an addition of A and B's complement. By adding 1 through the carry input, it is possible to implement 2's complement addition. In this diagram, addition of one 2 to 1 multiplexer allows us to select either B input without or with complement. Binvert input value decides whether to perform addition or subtraction. It should be noted that the value of Binvert should be 0, to perform logical operation. Figure indicates the function table for 1-bit ALU.

The adder with multiplexer can help us to combine adder and subtractor circuit together in a single circuit. When binvert=0, the multiplexer will send B directly to the multiplexer which enables the addition. If Binvert=1 then inverted input will be taken and transferred to the adder which performs subtraction. From the design of 1-bit ALU and function table, it is clear that it possible to perform addition as well as subtraction through an ALU. Subtraction is implemented by setting initial carry as 1 which is normally 0 for addition. In short, there are 2 main changes in the system. First each B bit is inverted and initial carry is set to 1.

Table: Function table of 1-bit ALU.

S2	S1	So	Binvert	Cin	Output	Operation
0	0	0	0	0	Y=A'	Complement
0	0	1	0	0	Y=A. B	AND
0	1	0	0	0	Y=A + B	OR
0	1	1	0	0	Y= A ⊕ B	XOR
1	X	X	0	0	Y= A+B	Addition
1	X	X	0	1	Y= A+B+Cin	Add with carry
1	X	X	1	0	Y= A-B	Subtraction (1's compl. Addition)
1	X	X	1	1	Y= A –B+cin	2's complement addition

2.2n-bit ALU Design

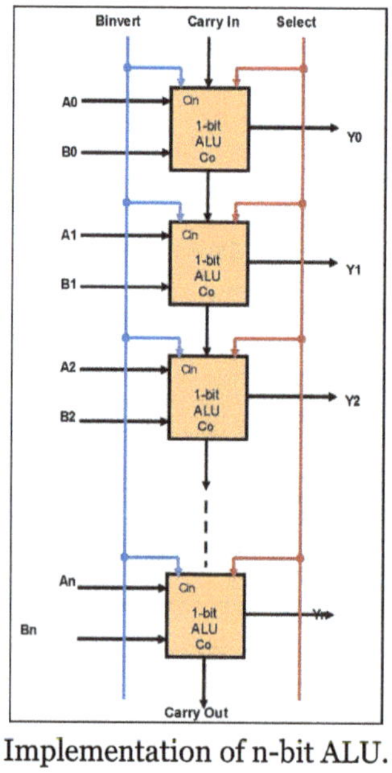

Implementation of n-bit ALU.

Once the complete 1-bit ALU is implemented using various combination block, we can extend the design to build n-bit ALU. Figure indicates cascading of 1-bit ALUs to construct n-bit ALU. One can thus extend the concept to 4-bit, 8-bit, 16-bit or 32-bit ALU. The input carry is applied to the first 1-bit ALU and the final carry is obtained from the last 1-bit ALU. Select lines and the B invert lines are connected in common. Select lines can decide on the nature of operation like arithmetic or logic operation. The binary value given to select lines along with Binvert and carry in will decide final function output as n-bit arithmetic or logic output. The function table for the n-bit ALU can also be prepared on the similar lines of 1-bit ALU.

ALU Design Process

The ALU design process begins with the specifications of ALU:

- Identify Arithmetic and Logical functions needs to be implemented.
- Start with word problem or the function table.
- Use divide and conquer approach i.e. Design arithmetic and logic unit separately.
- For arithmetic unit map the function table to an adder & build the input logic.
- For logic unit use basic logic gates and multiplexer for implementation.
- Integrate arithmetic and logic unit.

Example: Design a 4 bit ALU. Let us start with the functional specifications of the ALU. Table below indicates some of the different possible arithmetic operations. Following table indicates the select inputs and the desired functional output.

Table: Functional specifications of the ALU.

S2	S1	So	Output	Arithmetic operation
0	0	0	X	Transfer
0	0	1	X+1	Increment
0	1	0	X+Y	Add
0	1	1	X+Y+1	Add with carry
1	0	0	X+Y'	1's compl. Subtraction
1	0	1	X+Y'+1	2's compl. subtraction
1	1	0	X-1	Decrement
1	1	1	X	Transfer

Let us now map the function table to an adder. Table below indicates what the adder's inputs should be for each of our eight desired arithmetic operations. In this table, the adder input Ci is always the same as function select code bit So. The input A is always connected to X. The value of input B depended on function select inputs S2 and S1. The desired output of ALU depends on all these inputs.

Table: Logic and arithmetic functions.

S2	S1	B INPUT
0	0	0000
0	1	Y
1	0	Y'
1	1	1111

Function Select			Expected arithmetic function		Required Adder inputs		
S2	S1	S0			A	B	Ci
0	0	0	X	Transfer	X	0000	0
0	0	1	X+1	Increment	X	0000	1
0	1	0	X+Y	Add	X	Y	0
0	1	1	X+Y+1	Add with carry	X	Y	1
1	0	0	X+Y'	1's compl. Subtraction	X	Y'	0
1	0	1	X+Y'+1	2's compl. subtraction	X	Y'	1
1	1	0	X-1	Decrement	X	1111	0
1	1	1	X	Transfer	X	1111	1

As a next step let us build the input logic required for the B-input of an adder. Figure below indicates the desired inputs required for the input logic and the logic diagram of an ALU. Here, what is required is to compute input B given the arithmetic unit input Y and the functional select code.

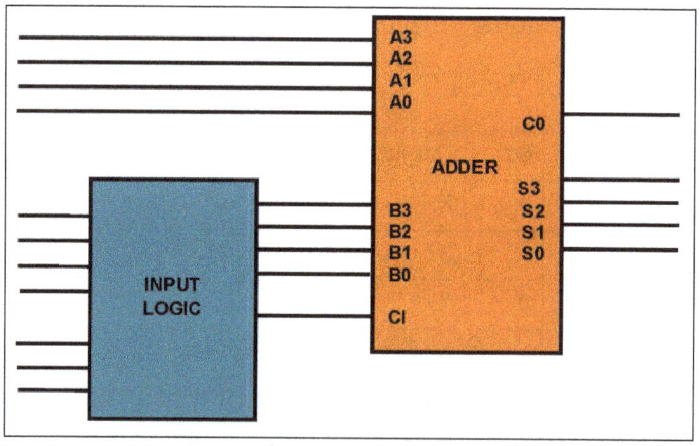

Input logic for Adder block.

The logic unit supports different logical functions on two multi-input X and Y producing the output Gi. The logic function table is shown in table below, indicates that there are four possible functions selected by logic select lines.

Table: Function table of Logic Unit.

S2	S1	Output
0	0	Gi=Xi.Yi

0	1	Gi=Xi+Yi
1	0	Gi=Xi⊕Yi
1	1	Gi=Xi

This table can be implemented using multiplexers and few basic gates. It should be noted that for n-bit logic unit we need a four to one multiplexers. Next step is to prepare combined the function tables of both arithmetic and logic functions together as shown in table below. Let us incorporate S3=0 for all arithmetic functions and logic functions when S3=1. There are already three variables in the arithmetic function table.

Table: Complete ALU function Select table.

Function select lines				
S3	S2	S1	S0	Operation
0	0	0	0	G=X
0	0	0	1	G=X+1
0	0	1	0	G=X+Y
0	0	1	1	G=X+Y+1
0	1	0	0	G=X+Y'
0	1	0	1	G=X+Y'+1
0	1	1	0	G=X-1
0	1	1	1	G=X
1	X	0	0	G=X and Y
1	X	0	1	G= X or Y
1	X	1	0	G= X xor Y
1	X	1	1	G=A'

The complete ALU diagram as shown in figure integrates both AU and LU. AU performs arithmetic functions and LU performs logic functions on 4-bit Xi and Yi decided by select inputs. S3 select input is connected to select line of quad 2 to 1 multiplexer which either selects AU or LU. For 4 bit operands, four 2 to 1 multiplexer is used. AU can also generate a carry output.

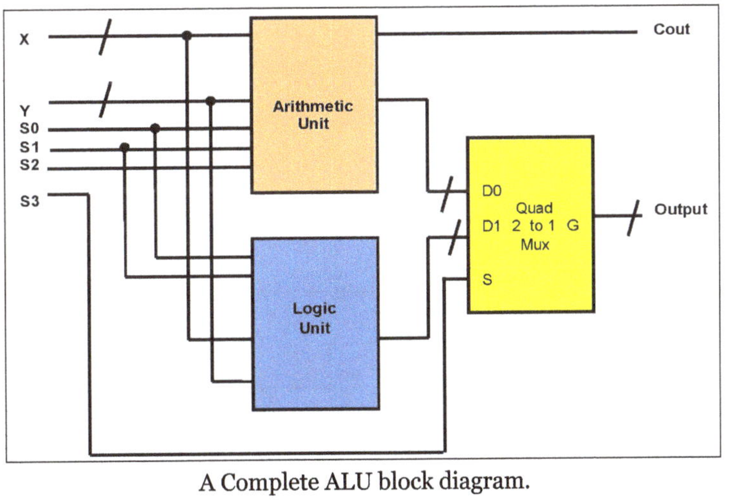

A Complete ALU block diagram.

ALU ICs

Arithmetic and logic unit is very popular device which is capable of performing different arithmetic and logical functions. Basically, ALU is the main part of central processing unit (CPU) of the computer. Let us study commercially available IC 74181. It is one of the popular and most widely used ALU. Figure indicates the logic symbol of IC74181.

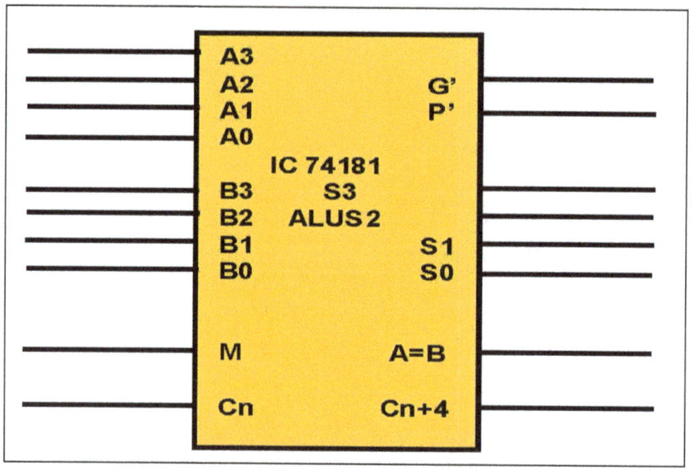

Functional diagram of IC 74181.

IC 74181 is basically a 4 bit ALU which can perform all the possible 16 logic operations on two operands and a variety of arithmetic operations. Features of IC74181:

- Arithmetic operating modes: Addition, subtraction, shift operand A one position, magnitude comparison and twelve other arithmetic operations.
- Logic function modes: XOR, AND, OR, NOR, NAND, comparator, and ten other logic operations.
- Full Look: Ahead logic for high speed operations on Long word.

Functional Description

The IC 74181 is an arithmetic logic unit (ALU). This ALU is capable of performing 16 binary arithmetic operations on two 4-bit words as shown in Table. These operations are selected by the four function select lines (S3, S2, S1, So) and include addition, subtraction, decrement, and straight transfer.

Mode Select Inputs				Active LOW inputs and outputs		Active HIGH inputs and outputs	
S3	S2	S1	So	Logic (M=H)	Arithmetic (M=L, Cn=L)	Logic (M=H)	Arithmetic (M=L, Cn=H)
0	0	0	0	A'	A minus 1	A'	A plus 1
0	0	0	1	(AB)'	AB minus 1	(A+B)'	(A+B) plus 1
0	0	1	0	A'+B	AB' minus 1	A'B	(A+B') plus 1

0	0	1	1	1	Minus 1	0	Zero
0	1	0	0	(A+B)'	A plus (A+B')	(AB)'	A plus AB' plus 1
0	1	0	1	B'	AB plus (A+B')	B'	(A+B) plus AB' plus 1
0	1	1	0	(A⊕B)'	A minus B minus 1	A⊕B	A minus B
0	1	1	1	A+B'	A+B'	AB'	AB'
1	0	0	0	A'B	A plus (A+B)	A'+B	A plus AB plus 1
1	0	0	1	A⊕B	A plus B	(A⊕B)'	A plus B plus 1
1	0	1	0	B	AB' plus (A+B)	B	(A+B') plus AB plus 1
1	0	1	1	A+B	A+B	AB	AB
1	1	0	0	0	A plus A*	1	A plus A plus 1
1	1	0	1	AB'	AB plus A	A+B'	(A+B) plus A plus 1
1	1	1	0	AB	AB' plus A	A+B	(A+B') plus A plus 1
1	1	1	1	A	A	A	A

The arithmetic and logic function can be selected by selecting mode, M. If M=1 then logic operations on the operands. Similarly, when M=0, the ALU performs the arithmetic operations. The operations on active low operands and active high operands are shown in figure. In addition there are two additional lines carry in and carry out provided for cascading 'n' numbers of ALU. The devices provides full internal carry look ahead and provides for either ripple carry between devices Cn+4 output or for carry look ahead between multiples ALU using signal carry propagate (P) and carry generate (G).

Digital System Design

Digital systems have created a digital revolution everywhere. Digital convergence leads to transformation and merging of various analog systems into digital world. The applications of digital systems have entered in all areas of human activity right from use of computers, electronic gadgets, house hold appliances, medical equipment to smart phones for internet access and e-purchases. One of the important reasons for this growth of digital electronics could be the use of Integrated Circuits (ICs) in all designs.

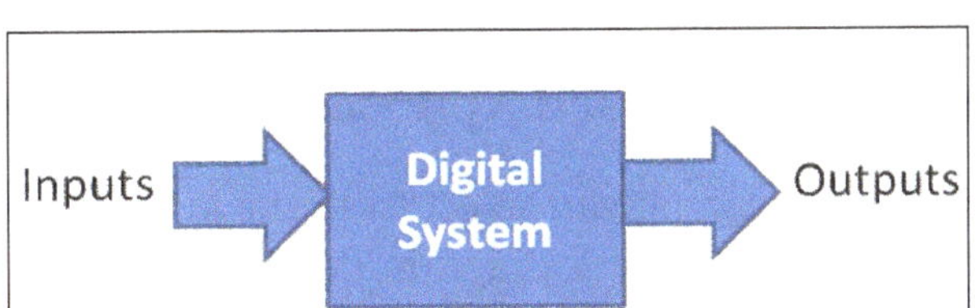

Typical digital system.

Digital design is the science of organizing information, defining operations, accepting inputs and generating outputs using discrete quantities as shown in figure. The quality of products, processes and services is increasingly dependent on digital system design. There are various techniques used for designing combinational and sequential circuits. In addition, one can select technology options i.e. TTL, CMOS, ECL etc. based on their relative advantages and disadvantages.

Traditionally, the digital system design methods were based on use of SSI and MSI circuits. These traditional methods of design are now challenged by recent digital devices called Programmable Logic Devices (PLDs), which are based on LSI circuits. The decision regarding the choice of design technique would depend upon several parameters like power requirements, size, cost, design complexity, frequency of operation etc.

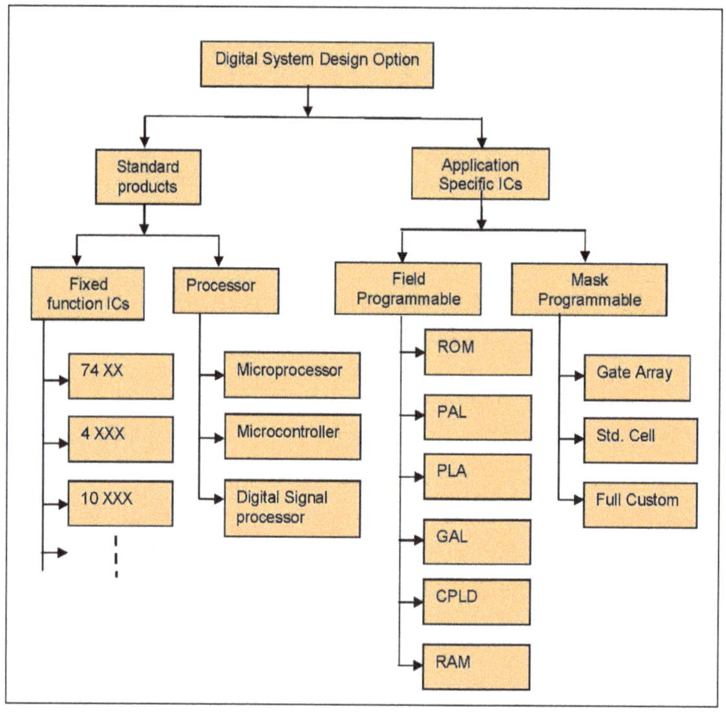

Digital System Design Options.

For designing the digital system for any application, the knowledge of various implementation technologies is essential. Designer must decide whether to use dedicated (fixed) functionality digital logic, or to use a software-programmed, processor-based system (designed based on a micropro-cessor, microcontroller, or digital signal processor, DSP), or to use an application specific ICs or hardware-configured programmable logic device (PLD), whether simple (SPLD), complex (CPLD), or the field programmable gate array (FPGA). Memory used for the storage of data and program code is integral to many digital circuits and systems.

Standard (Fixed Function) ICs based Approach

In traditional design approach, a designer can select from the available digital ICs most appropriate for the application, starting from a block diagram design concept and use available ICs. The advantages of traditional approach (i.e. fixed-function ICs) are:

- Shorter design cycle: Time require to design the system from concept to product is short.
- Lower development cost: As the design is based on fixed function ICs, the development cost is low.
- Easier to test & troubleshoot: As the designer is involved in designing the complete product, it is easy to test and troubleshoot.

Some of the disadvantages of traditional approach are:

- Larger space requirement: Fixed function ICs occupy finite space in the total board area.
- Higher power requirements: As there are discrete components in the device every IC would take finite power. The total required power is the sum of power consumption of each IC.
- Hardwiring for interconnections: Modifications are difficult. The complete design is based on discrete hardware. Their interconnections are hardwired on the board like PCB. On such board modification are difficult.
- Lesser reliability: Overall design involves many functional ICs along with separate interconnection the reliability interconnection is low.
- Lesser design flexibility: Due to complete hard-wiring in the design, there is less flexibility for the design.
- Difficult to Copyright: If one can carry out reverse engineering and trace the wiring diagram along with every IC, functionally it is very difficult.

Programmable ICs based Approach

There are two more approaches for designing the digital systems: Use of application specific integrated circuits (ASICs) usually mask programmable and use of programmable logic devices (PLDs) which are field programmable. All field programmable devices are programmed in laboratory. However, mask programmable devices must be sent to a manufacturer for implementation. The ASICs are developed to overcome the disadvantages of design based on fixed function ICs. In ASICs, single IC is designed and manufactured to implement logic functions. Many complex designs could be implemented easily using ASICs.

Advantages of ASICs

- Smaller space requirement (as integrated in single chip): The programmable ICs normally provide single chip solution hence the space requirement is small.
- Lower power requirements: Single chip solution has lower power consumption. Therefore overall power requirements are low.
- Large volume production can bring the cost down: If this design is used in large volume productive the cost of product can be brought down.
- Easier to copyright: One can copyright the design easily as it is a single chip solution. Major part of design is soft wired not hard wired.
- Highest reliability: The interconnection density is very low, this design offer highest reliability.

Disadvantages of ASICs

- Longer design and development cycle time: Usually ASICs are mask programmable hence the design and development cycle is longer and takes more time.

- Design and Development cost is high: As ASICs utilized specialized tools for the design and development, the total cost is higher.
- Special testing and verification tools are required: For all software based implementation there is need to have special testing and verification tools for different ASICs.

In the third approach, programmable logic devices may be used to implement digital systems. The PLDs have advantages of both the approaches as discussed above. Field programmable logic requires both hardware and software. PLDs are basically user configurable and re-programmable logic devices. User can program or change the program as per the requirement of the application. The main advantages are:

- Lower design and implementation cost: As programming done in the lab.
- Smaller space requirement (as integrated in single chip): The single chip solution on PLDs occupy smaller space on board.
- Lower power requirements: Due to single chip solution the power requirement are low.
- Higher reliability: As there are minimum interconnections on board, the reliability and design is higher.
- Easier to test: Usually simulation are carried out before programming it is easier to test.
- Easier to copy right: Single chip solution can easily have copyright for the design.
- More design flexibility: As PLDs reprogrammable and reconfigurable there is more design flexibility.

There are certain disadvantages of PLDs:

- Inefficient use of logic devices and resources: If the design involves less product term or sum terms then the AND & OR arrays are not utilize to their full capacity therefore, PLDs do not use all logic devices efficiently.
- Larger number of logic gates for implementation: PLDs provide all possible product terms using buffer/inverters at inputs hence there are large.

Comparison of Digital System Design Approaches

Digital systems can be implemented by either fixed function ICs or programmable ICs. Digital systems based on fixed function ICs make use of basic logic gates and others functions such as adder, multiplexers/demultiplexer, encoders/decoders, counters, shift registers. These ICs are designed by their manufacturer and produced in large quantities. On the other hand, programmable ICs are designed and can be easily configured by the user. Comparison between fixed function ICs and Programmable ICs is given in table.

Table: Comparison of Digital System design approaches.

Parameter	Fixed function ICs	Programmable ICs
Device Density	Less	More

Board Space	Large	Small
Power requirements	Large	Small
Development time	Less	More
Development cost	Low	High
Logic testing	Simple	Easy
Design complexity	Less	More
Design flexibility	Less	More
Design time	Less	Less
Design security	Low	High
Modification	Difficult	Easier

In order to compare digital system design approach, we will consider different parameters in first column for comparison. The second column describes the characteristics and fixed function IC and the third column describe the characteristics of programmable ICs design approach.

Permissions

All chapters in this book are published with permission under the Creative Commons Attribution Share Alike License or equivalent. Every chapter published in this book has been scrutinized by our experts. Their significance has been extensively debated. The topics covered herein carry significant information for a comprehensive understanding. They may even be implemented as practical applications or may be referred to as a beginning point for further studies.

We would like to thank the editorial team for lending their expertise to make the book truly unique. They have played a crucial role in the development of this book. Without their invaluable contributions this book wouldn't have been possible. They have made vital efforts to compile up to date information on the varied aspects of this subject to make this book a valuable addition to the collection of many professionals and students.

This book was conceptualized with the vision of imparting up-to-date and integrated information in this field. To ensure the same, a matchless editorial board was set up. Every individual on the board went through rigorous rounds of assessment to prove their worth. After which they invested a large part of their time researching and compiling the most relevant data for our readers.

The editorial board has been involved in producing this book since its inception. They have spent rigorous hours researching and exploring the diverse topics which have resulted in the successful publishing of this book. They have passed on their knowledge of decades through this book. To expedite this challenging task, the publisher supported the team at every step. A small team of assistant editors was also appointed to further simplify the editing procedure and attain best results for the readers.

Apart from the editorial board, the designing team has also invested a significant amount of their time in understanding the subject and creating the most relevant covers. They scrutinized every image to scout for the most suitable representation of the subject and create an appropriate cover for the book.

The publishing team has been an ardent support to the editorial, designing and production team. Their endless efforts to recruit the best for this project, has resulted in the accomplishment of this book. They are a veteran in the field of academics and their pool of knowledge is as vast as their experience in printing. Their expertise and guidance has proved useful at every step. Their uncompromising quality standards have made this book an exceptional effort. Their encouragement from time to time has been an inspiration for everyone.

The publisher and the editorial board hope that this book will prove to be a valuable piece of knowledge for students, practitioners and scholars across the globe.

Index